Advanced Memristor Modeling

Valeri Mladenov

Advanced Memristor Modeling

Memristor Circuits and Networks

MDPI • Basel • Beijing • Wuhan • Barcelona • Belgrade

MDPI

AUTHOR

Prof. Dr. Eng. Valeri Mladenov
Technical University of Sofia
Faculty of Automatics
Department of Theoretical Electrical Engineering
1000 Sofia, 8 St. Kliment Ohridski Blvd
Republic of Bulgaria

EDITORIAL OFFICE
MDPI
St. Alban-Anlage 66
Basel, Switzerland

For citation purposes, cite as indicated below:

Mladenov, V. *Advanced Memristor Modeling, Memristor Circuits and Networks Memristor Modeling, Memristor Devices, Circuits and Networks*; MDPI: Basel, Switzerland, 2019.

FIRST EDITION 2019

ISBN 978-3-03897-104-7 (Hbk)
ISBN 978-3-03897-103-0 (PDF)

doi:10.3390/books978-3-03842-103-0

Cover image courtesy of Valeri Mladenov.

Contents

CHAPTER I
Introduction to Memristors

CHAPTER II
Titanium Dioxide Memristor Models

CHAPTER III
Investigation of Memristor Circuits and Devices

CHAPTER IV
Analysis of Memristor Networks

About the Author

Valeri Mladenov received his PhD from Technical University of Sofia (TU Sofia), Bulgaria in 1993. In 2004, he became a Head of the Department of Theory of Electrical Engineering. Since June 2011, he has been Dean of the Faculty of Automation, since December 2011 he has been Vice Rector of TU Sofia, and since December 2015 he has been Director of the Directorate of Information and Public Relations. In 2014, he has been Deputy Minister of Education and Science in the Caretaker Government in Bulgaria. He is a guest lecturer at the Faculty of Electrical Engineering, Eindhoven University of Technology, The Netherlands, and many others. Dr. Mladenov's research interests are in the field of nonlinear circuits and systems, neural networks, artificial intelligence, applied mathematics, and signal processing. He has received many international research fellowships. He has more than 250 scientific papers in professional journals and conferences. He is a co-author of ten books and manuals for students. He has received many research grants from the Technical University of Sofia, Bulgarian Ministry of Education and Science, DAAD—Germany, NWO—The Netherlands, Royal Society—UK, NATO, TEMPUS, and others, and also with his team he has participated in many national and international projects—H2020, FP7, COST, Erasmus+ K1 and K2, DFG, and others. As a member of several editorial boards, Dr. Mladenov serves as a reviewer for a number of professional journals and conferences. He is a member of the International Neural Network Society (INNS), a member of the International Council of Large Electric Systems (SIGRE), a Senior Member of IEEE, a member of the IEEE Circuit and Systems Technical Committee on Cellular Nanoscale Networks and Array Computing, and Educational Activities Officer of the Bulgarian IEEE section. He is also a member of the Steering Committee of the International Symposium on Theoretical Electrical Engineering (ISTET), a member of the Management Boards of the Scientific and Technical Union of the Power Engineers, and a member of the Union of Automation and Informatics in Bulgaria.

Preface

The investigation of new memory schemes, neural networks, computer systems and many other improved electronic devices is very important for future generations of electronic circuits and for their widespread application in all the areas of industry. Relatedly, the analysis of new efficient and advanced electronic elements and circuits is an essential field of highly developed electrical and electronic engineering. The resistance-switching phenomenon, observed in many amorphous oxides, has been investigated since 1970 and is promising for inclusion in technologies for constructing new electronic memories. It has been established that such oxide materials have the ability to change their conductance in accordance to the applied voltage and memorizing their state for a long time interval. Similar behavior was predicted for the memristor element by Leon Chua in 1971. The memristor was proposed in accordance with symmetry considerations and the relationships between the four basic electric quantities—electric current i, voltage v, charge q and flux linkage Ψ. The memristor is a passive one-port element, together with the capacitor, inductor and resistor. The Williams Hewlett Packard (HP) research group has made a link between resistive switching devices and the memristor proposed by Chua. In addition, a number of scientific papers related to memristors and memristor devices have been issued and several models for them have been proposed. The memristor is a highly nonlinear component. It relates the electric charge q and the flux linkage Ψ, expressed as a time integral of the voltage v. It has the important capability of remembering the electric charge passing through its cross-section, and its respective resistance, when the electrical signals are switched off. Due to its nano-scale dimensions, non-volatility and memorizing properties, the memristor is a sound potential candidate for applications in high-density computer memories, artificial neural networks, and many other electronic devices.

A number of memristor models have been proposed in order to analyze their behavior in electric fields. Each model contains two basic equations. The first equation represents the relationship between the memristor voltage and current. This relationship is a state-dependent function—the state is related to the charge accumulated in its nanostructure. The second equation associates the time derivative of the memristor state variable and the current. The best models, such as the Biolek and Joglekar models, and the Boundary Condition Memristor model (BCM), describe titanium dioxide memristors and contain a window function in the right side of the state differential equation. The main window functions are state-dependent polynomials with a fixed positive integer exponent. The value of the applied exponent determines the nonlinearity of the used window function. The window function is used in the models to represent nonlinear ionic drift in the memristor nanostructure at high voltages. It has been established through many experiments and measurements that the nonlinearity of the ionic dopant drift depends on the applied voltage. However, in the existing basic memristor models, the integer exponent in the respective window function has a fixed value. By reason of the use of this constant integer exponent in the applied window function, the nonlinearity of the corresponding models does not depend on the memristor voltage. To the best of the author's knowledge, there is no established relationship between the positive integer exponent in the window functions applied in the basic memristor models and the memristor voltage. A motivating factor for the work presented here was to fill this gap and propose modifications to several titanium dioxide memristor models, and detailed analysis of the memristor element and several basic memristor circuits and devices. The modified models more realistically represent the voltage-dependent nonlinearity of the ionic dopant drift, according to the memristor state variable. The main modifications to the models are associated with the use of a voltage-dependent integer exponent in the window functions, for increasing the ionic motion nonlinearity, and a more realistic representation of the behavior in the general electric mode. This book presents a detailed analysis of several specific phenomena in the titanium dioxide memristor, such as the parasitic mutual inductances and capacitances in a memristor crossbar, the temperature influence on memristor behavior, and internal diffusion. The principle advantages of the proposed memristor models are the use of window functions with increased nonlinearity, which improve the memristor model behavior representation for higher voltages and avoid issues of lack of convergence.

This monograph summarizes results from several of the author's papers about memristors and memristor circuits. These papers are especially based on titanium dioxide memristors, which still have very wide application, and for this reason are main objects of investigation. It is organized as follows. Chapter 1 is an introduction to memristors; the basic physical description of the titanium dioxide memristor nanostructure and processes associated with the ionic current is presented. In Chapter 2, the fundamentals of titanium dioxide memristor modeling and simulations are described in detail, using the existing models and the models modified by the author. Chapter 3 reports the analysis of several basic memristor devices and circuits, such as memristor generators, integrators, and anti-parallel and series circuits, applying the author's memristor models. In Chapter 4, the investigation of memristor networks—memories and artificial neurons—is shown, paying attention to the basic electrical parameters and properties of the proposed memristor models and devices, and especially the modified memristor synaptic circuits. Finally, concluding remarks are presented.

VALERI MLADENOV
Technical University of Sofia
Sofia, Republic of Bulgaria

Acknowledgments

The author wants to thank to his assistant and collaborator Dr. Eng. Stoyan Kirilov for his valuable help during the technical preparation of the present manuscript.

This research is supported by national co-financing (contract No. ДКОСТ01/14) of COST Action No.IC1401 MemoCIS. Funding: The funding sponsors had no role in the design of the study; in the collection, analyses, or interpretation of data; in the writing of the manuscript, or in the decision to publish the results.

The author declares no conflict of interest. The sponsors had no role in the design of the study; in the collection, analyses, or interpretation of data; in the writing of the manuscript, and in the decision to publish the results.

Target Audience and Background Knowledge

This monograph was written for Ph.D. students, engineers and scientists, who want to enrich their knowledge about memristor modeling and investigation of memristors and memristor circuits, devices and networks. Readers should be familiar with the fundamentals of electronic and electrical engineering.

Numerical Methods

The solution of the basic differential equations and the describing equations, according to the fundamental laws of electrical engineering, were realized using the finite difference method, and the results were derived in MATLAB and OrCAD PSpice environment.

CHAPTER I

Introduction to Memristors

1.1. General Information for Memristors

The resistance-switching phenomenon, observed in a number of amorphous metal and transient chemical oxides, such as SiO_2, Al_2O_3, Ta_2O_5, has been investigated since 1970 [1]. It has been established that such oxide materials placed in a metal-oxide-metal structure have the capability of changing their conductance in accordance to the applied voltage and memorizing their conductance for a long-time interval [1,2]. Similar unusual and surprising behavior has been predicted for the memristor element by Leon Chua in 1971 [3].

The memristor element is proposed in accordance to symmetry considerations and the relations between the four basic electric quantities (current, voltage, electric charge and flux linkage) [3]. The electric charge is defined as a time integral of the current, and the magnetic flux is expressed as a time integral of the voltage. The resistor relates the voltage and current, the inductor is described with Faraday's Law using the relation between the current and magnetic flux, and the basic capacitor equation is expressed using the relation between electric charge and voltage. The proposed fourth fundamental two-terminal element relates the flux linkage, expressed as a time integral of the applied voltage, to the electric charge. The relations between the described electrical quantities and the fundamental one-port elements [4] are presented in Figure 1.

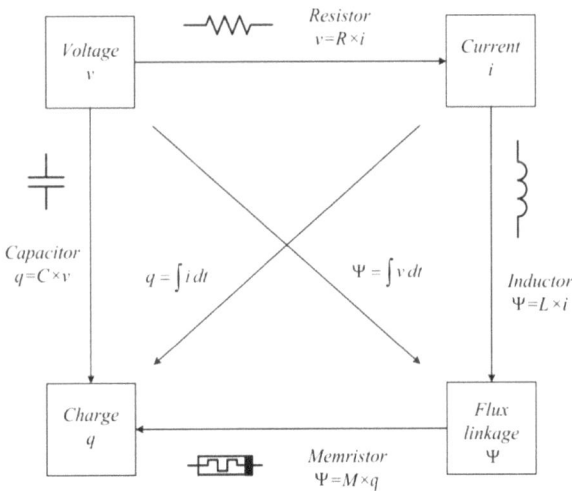

Figure 1. The relationships between the four basic electrical quantities (current i, voltage v, electric charge q and flux linkage Ψ) and the fundamental two-terminal elements (the resistor with a resistance R, the capacitor with a capacitance C, the inductor with an inductance L, and the memristor M). M is also used to denote the resistance of the memristor element, the so-called memristance.

The memristor is an essential passive one-port element together with the resistor, inductor, and capacitor [3–5]. The memristor is a highly nonlinear component [5,6]. It directly relates the electric charge and the flux linkage Ψ which is expressed as a time integral of the memristor voltage [3,4]. The memristor has the valuable capability of remembering the electric charge moving through its cross-section and its resistance M, when the electrical signals are switched off [4,5]. Its current–voltage relationship that will be discussed in detail later is a pinched hysteresis loop, of which the shape and range depend on both the magnitude and the frequency of the applied signal [4,6]. Since the memristor element could remember its conductance after the source is turned off, then the memristor could be applied as a non-volatile memory element [4–7]. The memory effect is based on accumulating electric charges in the memristor structure and holding them when the memristor voltage is zero.

1.2. Main Types of Memristors

Several basic types of memristors exist. They are based on different chemical and physical structures and have different principles of operation.

1.2.1. Titanium Dioxide Memristor

It is one of the leading nanostructured elements which still have broaden applications, owing to the fact that it is an object investigated in detail in the present technical research [4].

1.2.2. Polymeric Memristors

The polymeric memristors [8] are based on unique plastic materials. Single or parcels of molecules are able to conduct and switch currents and memorize information using charge accumulation. Instead of coding "0" and "1" as the amounts of charge stored in a silicon memory unit, polymer-based resistive random-access memory supplies information in a diverse way, for instance, based on the low or high conductance in reply to an applied external electrical field. The conductance states can be read non-destructively. Because electric conductivity is the multiplication of charge volumetric concentration and ionic mobility, changes in either the charge concentration or ionic mobility, or both can cause changes in the element resistance states. In the polymeric memory materials, the resistance bi-stability may appear from alterations of the characteristics of the switching material, and a few switching algorithms, together with charge transport, structure alteration and reduction-oxidation exchange [8].

1.2.3. Ferroelectric Memristors

Resistance switching has been observed in special sandwich-like structures consisting of ferroelectric thin films [9]. Their resistance can be reversibly changed between OFF state and ON state by applying an external signal. These two conductance states can be used in bistable memory components. More significantly, continuously tunable resistors, i.e., memristors, have been illustrated in ferroelectric tunnel junctions [9]. Resistance-switching behavior in $BiFeO_3$ has received enormous attention because it suggests additional degrees of independence for flexible devices. The established resistive-switching phenomena in such thin films appear to differ from those in the processing atmosphere and microstructures of the thin materials, given that several mechanisms could be involved. Since the polarization transfer has great stability for a chemical alteration and is basically rapid, the interfacial resistance-switching phenomenon is optimistic to the new ferroelectric memristors [9].

1.2.4. Spintronic Memristors

The spintronic memristor device [10] is based on an oxide thin film placed among two regions with magnetic properties. The first magnetic layer has permanent magnetic characteristics. The other film is a region with a field barrier. The magnetization of the free region on the left surface of the area divider is understood to be associated with the magnetic properties of the first region. The right surface magnetization is in the opposite location to the magnetic field of the first region. The domain barrier moves if the current flows through the described structure. This motion depends on the current path. The area wall can be fully transported to the left region and thus a total antiparallel magnetization between the regions occurs. This phenomenon leads to low conductance [10]. If the barrier is completely transported to the second region, a parallel magnetization between the first magnetic film and the pinned regions is realized. Then, high conductance is established. The state variable describes the length of the parallel magnetized section of the second region. This section has high conductivity and the antiparallel layer has lower conductance [10].

Principally, all types of memristor elements could be described using an equation describing the current–voltage relationship and an equation relating the time derivative of the state variable and the memristor current. In the next section, the titanium dioxide memristor nanostructure will be discussed in detail.

1.3. Basic Principles of Titanium Dioxide Memristor Nanostructure, and Its Physical Description in Electric Fields and Memory Effect

1.3.1. Main Memristor Nanostructure and Its Description

The titanium dioxide memristor nanostructure has been invented by Stanley Williams during his investigations on resistance-switching materials in Hewlett-Packard (HP) research group [4]. A simplified schematic of the memristor structure is presented in Figure 2(a). The electrodes of the memristor element are made of platinum or titanium material. The structure of the memristor element is based on a thin film of amorphous titanium dioxide material. The first region of the memristor element is formed in the titanium dioxide structure by doping with oxygen vacancies and the second one is based on pure titanium dioxide [4]. The oxygen vacancies have positive electric charge. The process of doping with oxygen vacancies is called electroforming and is based on applying a high direct current (DC) voltage with a value of 5 V [4]. Due to the high electric field, a partial evaporation of oxygen near the positive electrode occurs and a certain amount of oxygen vacancies appears in the left region of the titanium dioxide memristor sample [4]. The stoichiometric chemical formula of the doped region is TiO_{2-z}, where the index z has a value between 0.02 and 0.05.

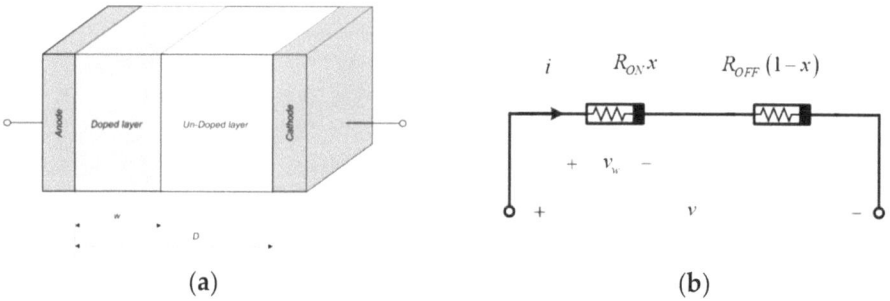

(a) (b)

Figure 2. (a) A simplified memristor nanostructure based on titanium dioxide; (b) A simplified electrical circuit substituted for the schematic of memristor nanostructure, using a series connection of two state-dependent nonlinear resistors, which represent the resistances of the doped and the undoped memristor layers, respectively.

The doped region of the memristor nanostructure has high conductance, while the pure titanium dioxide layer possesses very low conductance [4]. The length of the whole memristor nanostructure D is about 10 nm. The length of the doped region could be changed by applying an external electric field [4]. When a positive voltage is applied to the memristor, the positive potential on the anode repels the oxygen vacancies, and these oxygen vacancies start moving to the cathode. The length of the doped region increases until the oxygen vacancies reach the cathode. If a negative voltage is applied to the memristor nanostructure, then the anode has a negative potential. It attracts the oxygen vacancies and the length of the doped region decreases [4]. The resistances of the doped and undoped layers of the memristor depend on their instantaneous lengths, and on the specific resistances of these regions [4]. If the boundary between the doped and the undoped regions of the memristor element reaches the right edge of the whole memristor nanostructure, the doped region has a maximum length. In this state, the memristor has a minimum resistance, which has a value of 100 Ω [4]. This state of the memristor element is known as a fully closed state and the corresponding resistance is known as the ON-resistance (R_{ON}). If the boundary between the doped and the undoped regions of the memristor is on the left edge of the whole memristor nanostructure, then the whole region of the memristor is composed of pure titanium dioxide, and the memristor has a maximum resistance with a value of 16,000 Ω [4].

1.3.2. Mathematical Description of the Titanium Dioxide Memristor

This state of the memristor element, in which its maximum resistance is reached, is known as a fully open state, and its maximum resistance is known as the OFF-resistance (R_{OFF}) [4]. The instantaneous length of the doped layer is denoted with w and the whole length of the memristor is represented as D [4]. The state variable of the memristor x is expressed as a ratio between the length of the doped region w and the length of the whole memristor D [4]:

$$x = \frac{w}{D} \tag{1.1}$$

The doped and the undoped regions of the memristor element are connected in a series connection [4]. A simplified electrical circuit of the titanium dioxide memristor substituted for the schematic of its nanostructure is shown in Figure 2(b) to explain the memristor behavior in electric fields and further derive the state-dependent current–voltage relationship.

The resistance of the memristor region is proportional to the layer length and the specific resistance of the considered layer. The conductance of the saturated region is reversely proportional to the multiplication of the ON-resistance and its corresponding length.

The resistance of the undoped region is proportional to the length of the undoped region and the OFF-resistance of the element [4]. The equivalent resistance of the memristor R_{eq}, according to the electrical circuit (Figure 2(b)), is [4]:

$$R_{eq} = R_{ON}x + R_{OFF}(1 - x) \tag{1.2}$$

It is clear from Equation (1.2) that the state variable x determines the total resistance R_{eq} of the memristor element and the ON- and OFF-resistances. If a voltage signal v is applied to the element, the relationship between the memristor current i and the applied voltage v is expressed using the state-dependent Ohm's Law [4]:

$$v = i R_{eq} = i \left[R_{ON}x + R_{OFF}(1 - x) \right] \tag{1.3}$$

The voltage across the saturated layer of the element v_w is expressed using the Ohm's Law; it is a multiplication of the memristor current i, the resistance of the doped layer R_{ON} and the state variable x [4,11]:

$$v_w = i\,R_{ON}x = i\,R_{ON}\frac{w}{D} \tag{1.4}$$

The electric field strength in the doped region of the memristor $\vec{E}_w$ can be expressed as a ratio between the voltage across the doped layer v_w and its instantaneous length w [4,11,12]:

$$\vec{E}_w = \frac{v_w}{w} = \frac{i\,R_{ON}\frac{w}{D}}{w} = i\,\frac{R_{ON}}{D} \tag{1.5}$$

The relationship between the velocity of the oxygen vacancies $\vec{v}$, expressed as a time derivative of the instantaneous value of the doped region length w, and the electrical field strength in the saturated region $\vec{E}_w$ is:

$$\vec{v} = \frac{dw}{dt} = \frac{d}{dt}(x\,D) = D\frac{dx}{dt} = \mu\,\vec{E}_w = \mu\,i\,\frac{R_{ON}}{D} \tag{1.6}$$

where $\mu = 1 \times 10^{-14}\ \mathrm{m^2/(V{\cdot}s)}$ is the ionic mobility of oxygen vacancies in the titanium dioxide material [4]. Equation (1.6) can be transformed as follows [4]:

$$\frac{dx}{dt} = \mu\,\frac{R_{ON}}{D^2}\,i = k\,i \tag{1.7}$$

where k is a constant, dependent only on the memristor parameters μ, R_{ON} and D [4]. Equation (1.7) is the basic state differential equation of the idealized memristor element with linear ionic drift [4]. Equations (1.3) and (1.7) completely describe the memristor element behavior in an electric field [4] and are combined to constitute System of Equations (1.8):

$$\left|\begin{array}{l} v = i\,[R_{ON}x + R_{OFF}(1 - x)] \\[2mm] \frac{dx}{dt} = k\,i \end{array}\right. \tag{1.8}$$

If the applied memristor voltage v is a known variable, then the unknown variables are the memristor state variable x and the current i. First, the current i has to be expressed by the first equation in System of Equations (1.8) and then to be substituted into the state differential equation—the second equation in System of Equations (1.8). Therefore, System of Equations (1.8) can be rewritten as:

$$\left|\begin{array}{l} i = \frac{v}{[R_{ON}x + R_{OFF}(1-x)]} \\[2mm] \frac{dx}{dt} = ki = k\frac{v}{[R_{ON}x + R_{OFF}(1-x)]} \end{array}\right.$$

(1.9)

After performing the derivative of the state variable x of the element with respect to the time variable t in the second equation in System of Equations (1.9), the following state differential equation of the memristor element is acquired:

$$[(R_{ON} - R_{OFF})x + R_{OFF}]dx = kv\,dt \tag{1.10}$$

The left side of Equation (1.10) is integrated with respect to the state variable x from the initial value x_0 to the present value x, and the right side of the equation is integrated with respect to the time variable t from zero to the present moment of the analysis t, which can be described as:

$$\int_{x_0}^{x} [(R_{ON} - R_{OFF})y + R_{OFF}]\,dy = k\int_{0}^{t} v\,d\tau = k\Psi \tag{1.11}$$

where Ψ is the flux linkage. The state variable x of the memristor is limited according to physical considerations [4]. Its minimum value is zero when the border between the doped and the undoped regions is on the left edge of the whole memristor nanostructure, and the maximum value is unity when the described boundary between the memristor layers is on the right edge of the whole memristor nanostructure.

The initial value of the state variable x_0 is in the same region—[0, 1]. After integration of the state differential equation (1.11), the dependence between the state variable x and the flux linkage Ψ is derived—Equations (1.12a), (1.12b) and (1.2c):

$$(R_{ON} - R_{OFF})\frac{y^2}{2}\Big|_{x_0}^{x} + R_{OFF}y\Big|_{x_0}^{x} = k\Psi$$

$$\frac{(R_{ON} - R_{OFF})}{2}(x^2 - x_0^2) + R_{OFF}(x - x_0) = k\Psi$$

(1.12a)

$$\frac{(R_{ON} - R_{OFF})}{2}x^2 + R_{OFF}x - \left[k\Psi + \frac{(R_{ON} - R_{OFF})}{2}x_0^2 + R_{OFF}x_0\right] = 0$$

$$D = R_{OFF}^2 - 4\frac{(R_{ON} - R_{OFF})}{2}\left[k\Psi + \frac{(R_{ON} - R_{OFF})}{2}x_0^2 + R_{OFF}x_0\right]$$

The Equation (1.12) has two solutions, expressed with the next formulae—(1.13b) and (1.12c):

$$x_1 = \frac{-R_{OFF} + \sqrt{R_{OFF}^2 - 4\frac{(R_{ON}-R_{OFF})}{2}\left[k\Psi + \frac{(R_{ON}-R_{OFF})}{2}x_0^2 + R_{OFF}x_0\right]}}{R_{ON} - R_{OFF}} \qquad (1.12b)$$

$$x_2 = \frac{-R_{OFF} - \sqrt{R_{OFF}^2 - 4\frac{(R_{ON}-R_{OFF})}{2}\left[k\Psi + \frac{(R_{ON}-R_{OFF})}{2}x_0^2 + R_{OFF}x_0\right]}}{R_{ON} - R_{OFF}} \qquad (1.12c)$$

Only the second solution (1.14) of the state Equation (1.12) has a physical meaning, because the first one, expressed by (1.13), is higher than the upper limit of the physical value of the memristor state variable x. If it is assumed that $R_{ON} \ll R_{OFF}$, then the difference between the ON- and OFF-resistances could be approximately expressed as: $R_{ON} - R_{OFF} \approx -R_{OFF}$.

1.3.3. State–Flux Relationship of the Memristor

After the described simplification, using this approximation about the difference of the ON- and OFF-resistances of the memristor, the state–flux relationship of the considered element can be expressed as Equation (1.13):

$$\begin{aligned} x &= 1 - \sqrt{1 + \frac{2k}{R_{OFF}}\int_0^t v\,d\tau - x_0^2 + 2x_0} \\ x &= 1 - \sqrt{1 + \frac{2k}{R_{OFF}}\Psi - x_0^2 + 2x_0} \end{aligned} \qquad (1.13)$$

The first equation in System of Equations (1.13) confirms mathematically the memory effect of the titanium dioxide memristor element. It is observable that the memristor state variable x depends on the definite integral of the memristor voltage, and if the applied voltage becomes zero for a given time interval, the memristor state variable x retains its previous value, acquired immediately before the memristor voltage is switched off.

1.3.4. Current–Voltage Characteristic of the Memristor

Substituting the first equation in System of Equations (1.13) into the first equation in System of Equations (1.8), the current–voltage relationship of the memristor element is acquired—Equation (1.14):

$$i = \frac{v}{\left[(R_{ON} - R_{OFF})\left[1 - \sqrt{1 + \frac{2k}{R_{OFF}}\int_0^t v\,d\tau - x_0^2 + 2x_0}\right] + R_{OFF}\right]} \qquad (1.14)$$

10

1.3.5. Basic Memristor Characteristics

The derived state–magnetic-flux and current–voltage relationships in this case are valid for an idealized and simplified memristor element model. Several basic memristor models, based on physical experiments and data, will be discussed in the next chapter. Using System of Equations (1.13) and (1.14), the following basic results are obtained in MATrix LABoratory (MATLAB) Environment [13]. First, the described memristor element is investigated for an applied sinusoidal voltage signal, using System of Equations (1.13) and (1.14). The derived time diagrams of memristor voltage and current according to Equation (1.14) are presented in Figure 3(a). A sine-wave voltage signal with amplitude of 1 V and a frequency of 1 Hz is applied to the memristor element. The current flowing through the memristor i has a non-sinusoidal waveform. This confirms the nonlinear behavior of the considered element. The corresponding time graphs of memristor state variable x and flux linkage Ψ are shown in Figure 3(b). It is obvious that if the applied voltage v and the magnetic flux Ψ are sinusoidal functions, then the curves describing the memristor current i and state variable x are non-sinusoidal.

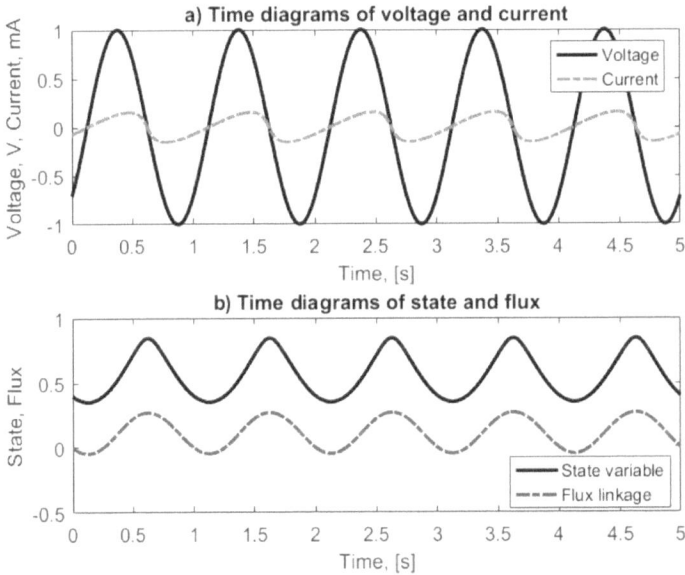

Figure 3. (a) Time diagrams of the applied memristor voltage $v = 1 \times \sin\left(2\pi t - \frac{\pi}{6}\right)$ and the current i in sinusoidal mode; **(b)** time diagrams of the state variable x and the flux linkage Ψ in sinusoidal mode.

These nonlinearities are due to the nonlinear state–magnetic-flux relationship of the memristor element. The state variable of the memristor x does not achieve its limiting values—zero and unity. The state–flux and current–voltage relationships are presented in Figure 4(a,b) for the description of the basic memristor characteristics and properties. The state–magnetic-flux relationship is a monotonically increasing nonlinear curve, while the current–voltage relationship is a pinched hysteresis loop. The state–magnetic-flux relationship is a single-valued curve. It is obvious that the current–voltage characteristic has several regions with negative differential resistance [4]. In these regions, the memristor current i increases when the voltage v decreases [4]. There are two basic branches in the current–voltage relationship for memristor ON and OFF states, respectively [4].

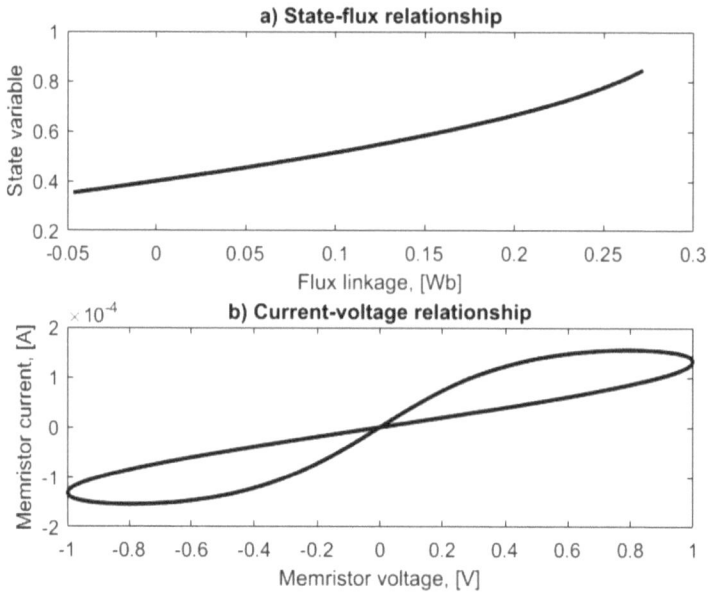

Figure 4. (a) The corresponding state-flux and (b) current–voltage characteristics of the memristor element in sine-wave regime, describing the main memristor properties.

The intersection of the main branches of the current–voltage characteristic of the element matches the origin of the current–voltage coordinate system. It could be concluded that the memristor is a passive element, although its current–voltage relationship has several regions with negative differential resistance [4].

1.3.6. *Memory Effect of the Memristor Element in Electric Fields*

This effect of the memristor element, operating in electric fields is visually expressed in Figure 5(a,b). The applied memristor voltage in this case is a sequence of positive rectangular pulses. The duty cycle of the pulse sequence is 50%. When a positive voltage pulse is applied to the memristor element, its state variable x increases, and the memristance (known as the resistance of the memristor element), proportional to the state variable x, decreases, according to Equation (1.2).

In the duration of the pauses between the voltage pulses, when the memristor voltage is zero, the state variable x of the memristor retains its previous value, and on the time diagram, it is presented with a horizontal line segments. The observed phenomenon confirms the memorizing effect of the memristor element. When a new positive voltage pulse is applied, the state variable increases again from its previously reached value. The state–flux and current–voltage relationships in pulse mode are presented in Figure 6(a,b). The passivity of the memristor is observed in Figure 6(b); the memristor current i is zero for a voltage signal v with a value of zero.

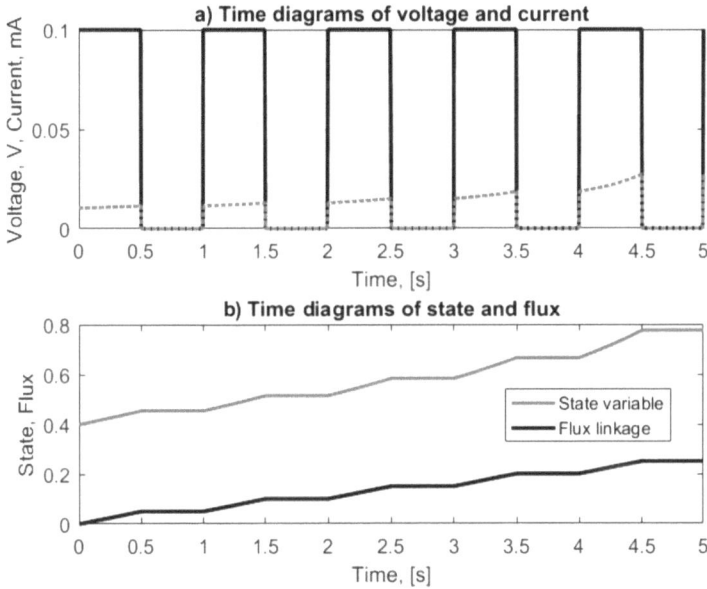

Figure 5. (a) Time diagrams of the applied memristor pulse voltage with an amplitude of 0.1 V, a main frequency of 1 Hz and a duty cycle of 50%, and the memristor current i; (b) time diagrams of the state variable x and the flux linkage $Ψ$ in pulse mode, visually presenting the memristor memory effect.

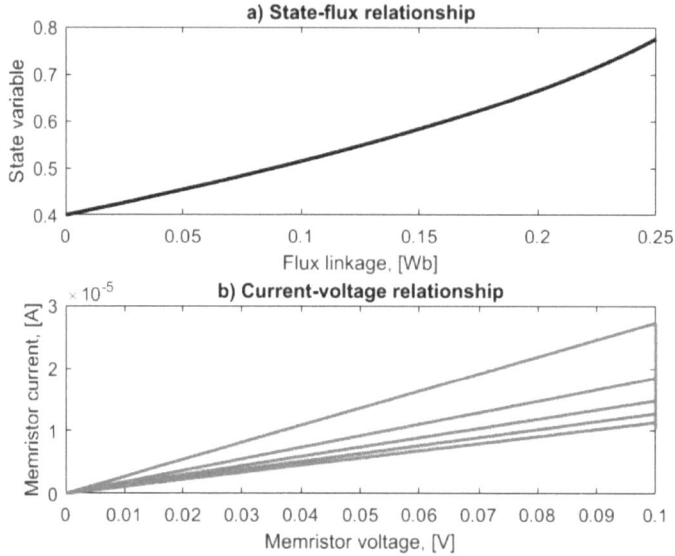

Figure 6. (a) The corresponding state–flux and (b) current–voltage relationships of the memristor element in pulse mode, illustrating the passivity of the memristor element. If the applied voltage v is zero, the memristor current i is also zero.

The described properties and characteristics of the memristor element are valid for the idealized case, when the state variable x does not reach its limiting values, according to the assumption of linear ionic dopant drift [4].

In the next chapters, several additional effects, associated with experimentally established nonlinear ionic drift [5,6], will be investigated and discussed. The inspiration for the present investigation is related to several established results by the author, lacking specific relationships and additional effects in the main memristor models. Therefore, the main purpose of the book is to provide a detailed description of the relationships between the nonlinearity of the ionic dopant drift in the titanium dioxide memristor and the applied voltage v, and to propose several improved memristor models and analyses of memristor devices and networks, based on the proposed memristor models.

In Chapter 2, the basic existing memristor models and the proposed improved models by the author are represented and discussed. In Chapter 3, the investigation of several memristor circuits and devices with the use of the proposed advanced memristor models by the author is presented. Chapter 4 represents the analysis of several memristor networks, such as memory crossbars and artificial neurons, with the application of the proposed memristor models.

CHAPTER II

Titanium Dioxide Memristor Models

2.1. Classical Memristor Models

The representation of memristor behavior in electric fields by System of Equations (1.8), describing the memristor state differential equation and the corresponding state-dependent Ohm's Law, has a certain disadvantage. Based on the analysis, if the voltage signal applied to the memristor is at a high level or the corresponding signal frequency is very low, it is mathematically possible for the state variable x to become higher than unity or lower than zero. However, this behavior is impossible by physical considerations [3,4], so a modification of System of Equations (1.8) has to be applied. The easiest way of limiting the state variable x is the application of a window function with the following properties: for the limiting values of the state variable—zero and unity, the window function has to be equal to zero; for the other values of the state variable between zero and unity, the window function is near unity, or between zero and unity [5,6]. The modified state differential equation of the memristor in a combination with the state-dependent Ohm's Law is then described as System of Equations (2.1) [5,6,11]:

$$\frac{dx}{dt} = k\,i\,f(x)$$
$$v = i\left[R_{ON}x + R_{OFF}(1-x)\right] \tag{2.1}$$

where $f(x)$ is the applied window function [5,6,11]. The first equation in System of Equations (1.8) remains the same as the second equation in System of Equations (2.1), which fully describes the corresponding memristor model [5,6,11]. The solution to System of Equations (2.1) for voltage-supplied memristor is based on the memristor current expressed by the second equation in System of Equations (2.1) and substitution of the current into the first equation in System of Equations (2.1), which can be shown as follows in Equation (2.2):

$$i = \frac{v}{[R_{ON}x + R_{OFF}(1-x)]}$$
$$\frac{dx}{dt} = k\,i\,f(x) = k\,\frac{v}{[R_{ON}x + R_{OFF}(1-x)]}\,f(x) \tag{2.2}$$

After separating the state variable x and the time variable t, the following state differential Equation (2.3) for the memristor element is derived:

$$\frac{[R_{ON}x + R_{OFF}(1-x)]}{f(x)}\,dx = k\,v\,dt \tag{2.3}$$

If the left side of Equation (2.3) is too difficult or impossible for analytical integration, the finite difference method could be applied for the numerical solution of this differential equation. The solution to Equation (2.2) relates the state variable x and the flux linkage Ψ which is expressed as a time integral of the voltage. The denoted window function above could be used for other types of memristors, if they could be described by similar equations to System of Equations (2.1).

17

2.1.1. Strukov and Williams's Memristor Model

During the investigation of the titanium dioxide memristor nanostructures, Strukov and Williams in Hewlett Packard (HP) research laboratory proposed a simple parabolic window function for the titanium dioxide memristor description [4], expressed by Equation (2.4):

$$f_{sw}(x) = x(1-x) \tag{2.4}$$

The proposed parabolic window function has high nonlinearity and is used for limiting the state variable and for introducing nonlinearity to the ionic dopant drift representation [4]. The maximum value of this window function is 0.25 [4]. The corresponding System of Equations (2.5) describes this memristor model [4]:

$$\frac{dx}{dt} = k\,i\,f_{sw}(x) = k\,i\,[x(1-x)]$$
$$v = i\,[R_{ON}x + R_{OFF}(1-x)] \tag{2.5}$$

The following approximation for the resistance of the memristor M (2.6) is applied [4,5], using the assumption that the ON-resistance is many times lower than the OFF-resistance of the memristor element:

$$M = R_{ON}x + R_{OFF}(1-x) \approx R_{OFF}(1-x) \tag{2.6}$$

This approximation is valid if the OFF-resistance of the memristor is many times higher than its ON-resistance. In this case, the first term in the expression above for the resistance of the memristor element could be neglected [4]. After substituting Equation (2.6) above into the first equation in System of Equations (2.4), the state differential equation of the memristor (2.7) is obtained [4]:

$$\frac{1}{x}dx = \frac{1}{R_{OFF}}k\,v\,dt \tag{2.7}$$

After integration of Equation (2.7) within the respective integration ranges for the memristor state variable x and the time variable t, the following state–flux relationship of the memristor, expressed by Equation (2.8) is derived:

$$x = x_0\exp\left(\frac{k}{R_{OFF}}\int_0^t v\,d\tau\right) = x_0\exp\left(\frac{k}{R_{OFF}}\Psi\right) \tag{2.8}$$

In this case, an exponential relationship between the state variable x and the flux linkage Ψ is obtained. Its shape is related to the initial value of the state variable x_0. The corresponding current–voltage relationship of the memristor element is derived after substitution of Equation (2.8) into the second equation into System of Equations (2.4). The following Equation (2.9) is derived:

$$i = \frac{v}{\left[(R_{ON} - R_{OFF}) x_0 \exp\left(\frac{k}{R_{OFF}} \int_0^t v \, d\tau \right) + R_{OFF} \right]} \tag{2.9}$$

In this case, the initial value of the state variable has to be higher than zero. If the initial value of the memristor state variable is chosen equal to zero, the other values of the state variable are also equal to zero, although the flux linkage is changing. This behavior of the state–flux relationship in this case does not physically describe the real state–flux relationship and it is derived according to the assumed approximation. The basic electric quantities related to memristor operation in the time domain are presented in Figure 7(a,b) for further discussion. The derived time graphs of the memristor voltage $v(t) = 3\sin(2\pi \times 0.5t - \frac{2\pi}{3})$, and the current i according to Strukov and Williams's memristor model are presented in Figure 7(a,b) for describing the memristor nonlinearity and visual expression of the basic quantities in the time domain.

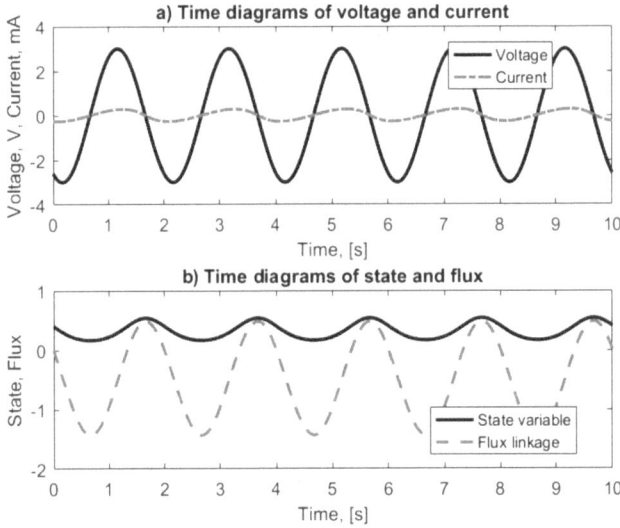

Figure 7. (a) Time diagrams of the applied sinusoidal memristor voltage $v(t) = 3\sin\left(2\pi \times 0.5t - \frac{2\pi}{3}\right)$ and the non-sinusoidal current i according to the Strukov and Williams's model with a parabolic window; (b) time diagrams of the memristor state variable x and flux linkage Ψ, expressed as a definite time integral of the applied voltage.

By comparing the results to these derived by the idealized memristor model without a window function, given in Figure 4, it could be concluded that the current in the present case has several times lower magnitude than that in the idealized case. This behavior of the Strukov and Williams's memristor model could be explained with the low maximum value of the applied parabolic window function, i.e., 0.25. The corresponding memristor state variable in the present case changes in a narrower range, because its change is proportional to the time integral of the applied voltage. The corresponding state–flux and current–voltage relationships of the memristor are shown in Figure 8(a,b), for further descriptions of these basic memristor characteristics. The state–flux relationship is a monotonically increasing nonlinear function. Its nonlinearity determines the area of the corresponding current–voltage characteristic in the respective coordinate system of the memristor element [3,4].

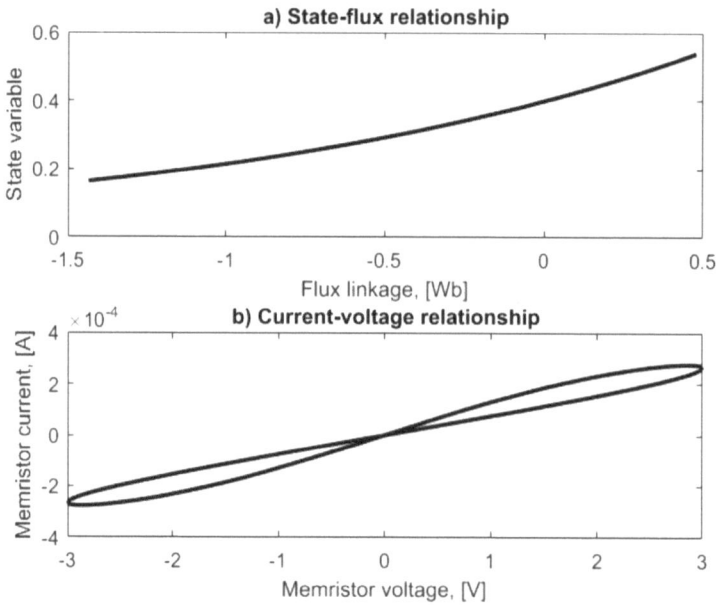

Figure 8. (**a**) The state–flux characteristic of the memristor element for a sinusoidal memristor voltage; and (**b**) the corresponding current–voltage characteristic of the memristor element.

It is confirmed by many experiments that if the curve describing the state–flux relationship is similar to a straight line, then the corresponding current–voltage relationship curve is also a straight line. A fragment of Strukov window function, derived based on the analysis for the same sinusoidal voltage signal, is presented in Figure 9(a) for illustrating the trajectory of the memristor operating point in the field of the coordinates representing the state variable x and the corresponding segment of the window function $f(x)$.

The presented fragment is only a part of the whole window function, which could be obtained for the hard-switching operation mode (for higher voltages and lower frequencies). The time diagram of the resistance of the memristor element (the memristance) is presented in Figure 9(b). It has a non-sinusoidal form and the range of the memristance is between 700 and 1350 Ohms.

The time graphs of the voltage and the corresponding state variable in pulse mode are given in Figure 10. The applied voltage is a sequence of positive and negative impulses. The maximum level of the memristor voltage is 0.3 V. In the time intervals, when positive and constant memristor voltages are used, the state variable x increases. The relationship between the state variable x and the time integral of the applied voltage is almost a linear function. When the applied voltage v has a negative polarity, the state variable x decreases and is proportional to the integral of the applied voltage.

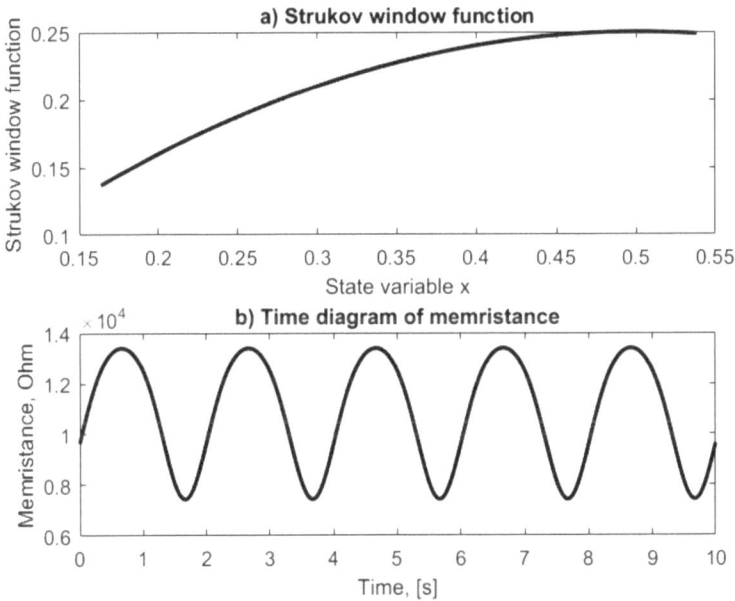

Figure 9. (a) A fragment of Strukov's window function, derived from the memristor operation; it represents only a part of the whole parabolic function; (b) time diagram of the resistance of the memristor element (memristance), which in this case represents a soft-switching operation mode.

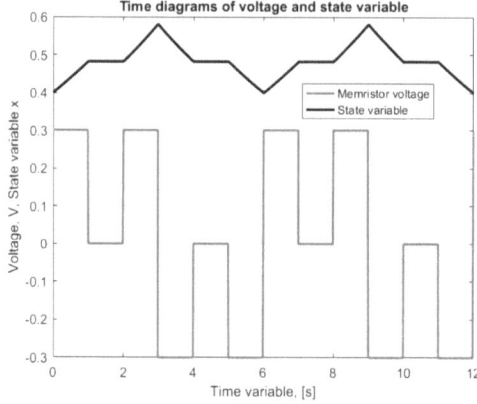

Figure 10. Time diagrams of the applied pulse voltage v and the corresponding state variable x for expression of the memristor memory effect. When the memristor voltage is zero, the state variable of the memristor holds its previous value.

The memory effect, associated with the memristor element, is observable. If the applied voltage is zero, the state variable retains its previous value. In these time intervals, the memristor state variable x is represented by a horizontal line, demonstrating that the state variable x is a constant quantity.

2.1.2. Joglekar Memristor Model

A frequently used polynomial window function with second and higher power in the memristor modeling was proposed by Joglekar [5]:

$$f_J(x) = 1 - (2x - 1)^{2p} \tag{2.10}$$

where p is a positive integer exponent [5]. The nonlinearities of the discussed window function and of the memristor ionic drift grow up with the decreasing of the exponent in the Joglekar window function [5]. Several Joglekar window functions for different integer exponents are presented in Figure 11 for visual representation of their different shapes and nonlinearities. The state differential equation of the memristor element, acquired by applying the Joglekar window function, in combination with the state-dependent Ohm's Law, is described as the following Equation (2.11) [4,5]:

$$\frac{dx}{dt} = k\,i\,f_J(x) = k\,i\,\left[1 - (2x - 1)^{2p}\right]$$
$$v = i\,[R_{ON}x + R_{OFF}(1 - x)] \tag{2.11}$$

The time diagrams of memristor voltage $v(t) = 3\sin(2\pi \times 0.5t - \frac{2\pi}{3})$, the current i, state variable x and flux linkage Ψ are presented in Figure 12(a,b). The applied memristor voltage signal v has a sinusoidal form, while the current i and the state variable x has highly non-sinusoidal form.

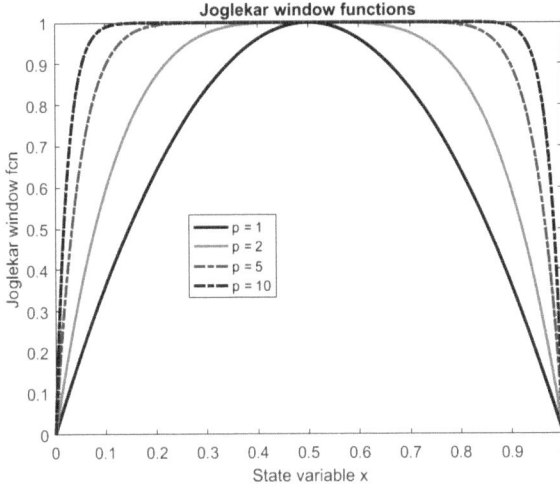

Figure 11. Joglekar window functions for different values of the integer exponent p, expressing their different shapes and nonlinearities. For lower integer exponent values, the nonlinearity of the window function is higher; with the increased positive integer exponent p in the Joglekar window function, the curves describing the Joglekar window function becomes almost linear, while only for the periphery, the Joglekar window function is nonlinear, approaching the abscissa axis.

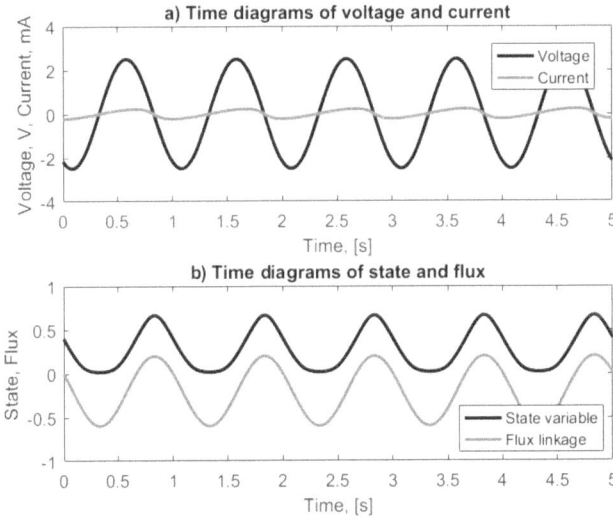

Figure 12. (a) Time diagrams of the applied sinusoidal memristor voltage $v(t) = 3\sin\left(2\pi \times 0.5t - \frac{2\pi}{3}\right)$ and the current i; (b) time diagrams of the state variable x and the flux linkage Ψ, according to Joglekar's model with a polynomial window function.

23

The state–flux and current–voltage characteristics of the memristor element, rendering to the Joglekar model, are given in Figure 13(a,b) for representation of the basic memristor characteristics and further discussion. The current–voltage pinched hysteresis loop has a larger area in the field of the respective coordinates, according to the corresponding results, derived by Strukov and Williams [4]. The corresponding time diagrams of the memristance (dependent on voltage, current, state variable and flux linkage) with the Joglekar window function are represented in Figure 14(a,b).

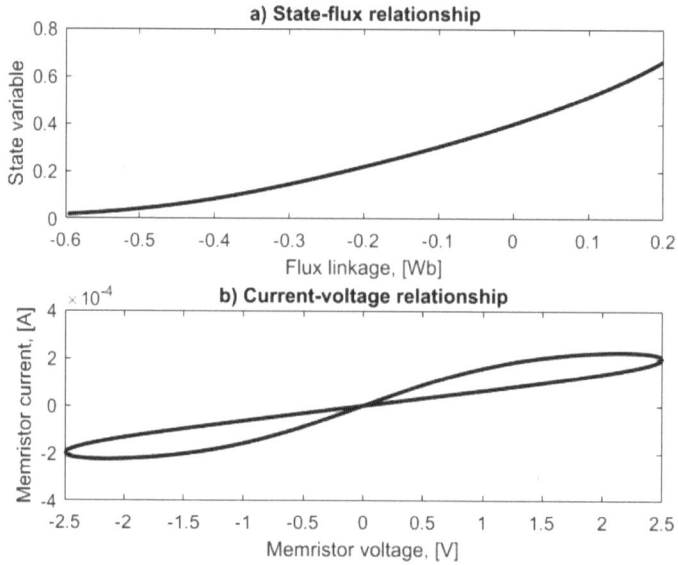

Figure 13. (a) The state–flux characteristic of the memristor element according to the Joglekar model, for the same sinusoidal memristor voltage as in Figure 12(a,b); (b) the current–voltage relationship of the memristor for the voltage $v(t) = 3\sin\left(2\pi \times 0.5t - \frac{2\pi}{3}\right)$.

The state variable x almost reaches its minimum value of zero and its corresponding maximum value is about 0.7. The flux linkage is a sinusoidal quantity, equal to the definite time integral of the voltage v [4,5]. After analyzing the state–flux relationship of the Strukov model, it could be concluded that, in the present case, this relationship has higher nonlinearity. The observed phenomenon could be explained with the higher maximum value of the Joglekar window function.

The presented segment of the window function has a large horizontal linear fragment because the used value of the positive integer exponent is comparatively high, i.e., $p = 5$.

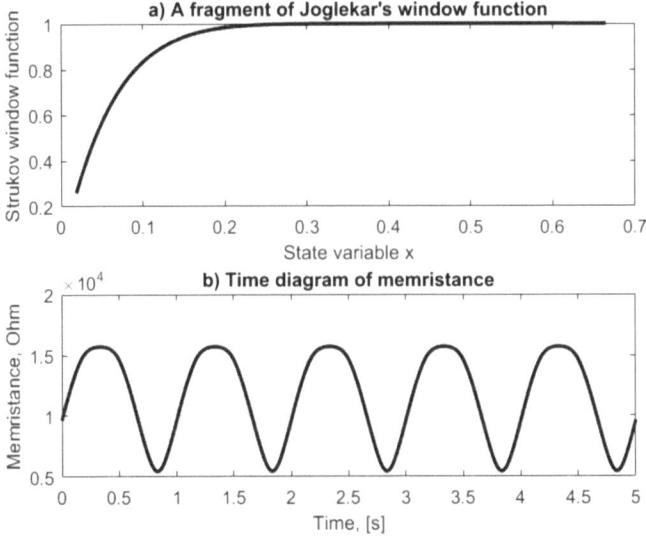

Figure 14. (a) A fragment of the Joglekar window function for $p = 5$; (b) time diagram of the resistance of the memristor (memristance) in the time domain in a soft-switching mode.

2.1.3. Biolek Memristor Model

A frequently used and widespread window function, dependent on the memristor state variable x and the current direction, was first proposed by Biolek [6] and is also identified as a Biolek window function, which can be written as the next Equation (2.12):

$$f_B(x, i) = 1 - [x - stp(-i)]^{2p} \tag{2.12}$$

where the used relay function "stp" is expressed by the following Equation (2.13) [6]:

$$stp(i) = \begin{cases} 1, & \text{if } i \geq 0 \quad (v \geq 0) \\ 0, & \text{if } i < 0 \quad (v < 0) \end{cases} \tag{2.13}$$

The Biolek memristor model is expressed with the next System of Equations (2.14) [4,6]:

$$
\begin{aligned}
&f_B(x, i) = 1 - [x - stp(-i)]^{2p} \\
&stp(i) = \begin{cases} 1, & \text{if } i \geq 0 \quad (v \geq 0) \\ 0, & \text{if } i < 0 \quad (v < 0) \end{cases} \\
&v = [R_{ON}x + (1 - x)R_{OFF}] \, i
\end{aligned}
\tag{2.14}
$$

where the last equation in System of Equations (2.14) is based on the state-dependent Ohm's Law for the titanium dioxide memristor element [4,6]. After substituting the second equation in System of Equations (2.14) into Equation (2.12), the following form of Biolek window function, written by Equation (2.15) is acquired:

$$f_B(x) = 1 - (x-1)^{2p}, \quad v(t) \le 0, \quad [i(t) \le 0]$$
$$f_B(x) = 1 - x^{2p}, \qquad\quad v(t) > 0, \quad [i(t) > 0]$$

(2.15)

Several Biolek window functions for different positive integer exponents are presented in Figure 15. Each Biolek window function is completed by two different branches, derived for different polarities of the memristor current. By increasing the value of the positive integer exponent p, the nonlinearity of Biolek window function decreases.

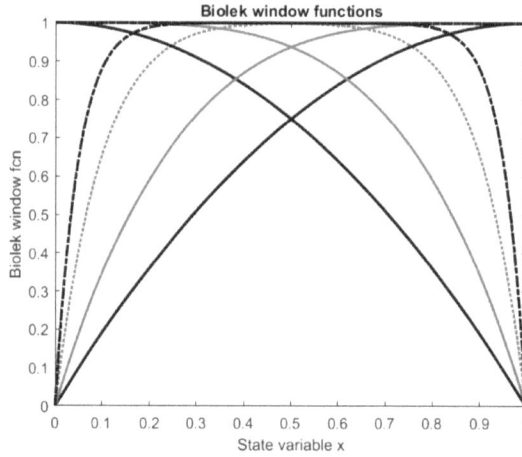

Biolek window functions

Figure 15. Biolek window functions for different values of the integer exponent p. With increasing of the positive integer exponent p in the model, the nonlinearity of the Biolek window function decreases.

The time diagrams of the applied sinusoidal memristor voltage $v(t) = 3\sin\left(2\pi \times 0.5t - \frac{2\pi}{3}\right)$, the corresponding current i, flux linkage Ψ and state variable x according to the Biolek model are presented in Figure 16(a,b). The current flowing through the memristor is rectified with respect to the applied voltage v. It is obvious that the memristor current i has a highly non-sinusoidal and non-symmetrical shape with the applied sinusoidal voltage. The state variable of the memristor x reaches its upper limiting value. Due to this phenomenon, the memristor operates in a state near the hard-switching regime. The corresponding state–flux and current–voltage characteristics of the memristor element according to the Biolek model in hard-switching mode are presented in Figure 17(a,b). The state–flux relationship is shown by a multi-valued hysteresis curve, according to the boundary effects. In this case, the curve describing the current–voltage relationship is asymmetrical, similar to that representing the current–voltage relationship of a semiconductor rectifier diode.

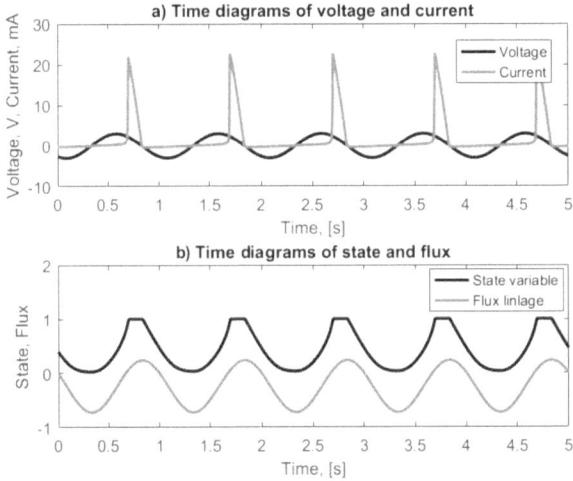

Figure 16. (**a**) Time graphs of the applied voltage $v(t) = 3\sin\left(2\pi \times 0.5t - \frac{2\pi}{3}\right)$ and the current i; (**b**) time diagrams of the memristor state variable x and flux linkage Ψ, according to the Biolek model; in the present case, the state variable x of the memristor almost reaches its limiting values.

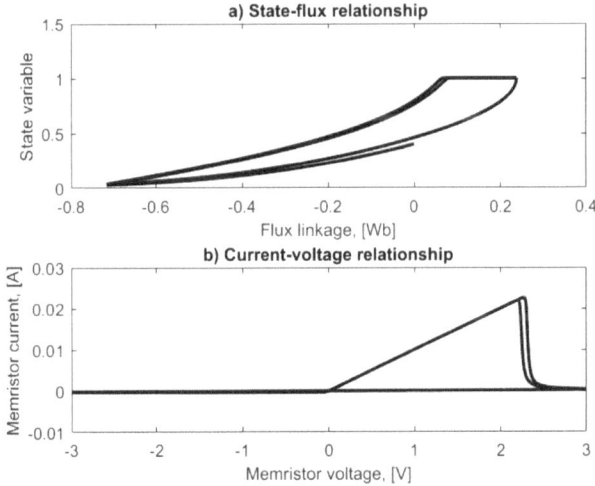

Figure 17. (**a**) The state–flux characteristics of the memristor element; and (**b**) the current–voltage relationship, according to the Biolek model. In the present case, the state–flux relationship is shown by a multi-valued hysteresis curve; a non-symmetrical curve shows the current–voltage relationship for the hard-switching mode.

27

A fragment of the applied Biolek window function derived in the operation process of the memristor element is presented in Figure 18(a). In this case, almost the whole window function could be observed because the memristor element operates in a state near a hard-switching mode. The corresponding time diagram of the memristance is given in Figure 18(b). The memristance in the present case is changing in a very wide range. This phenomenon corresponds to a hard-switching mode.

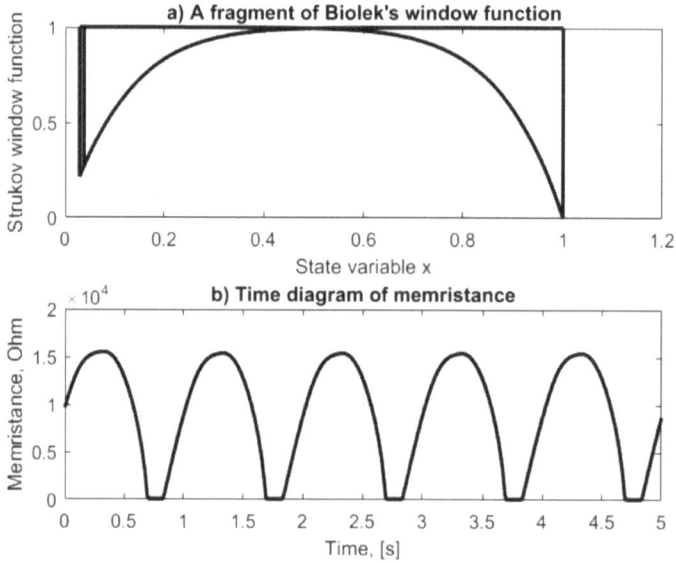

Figure 18. (a) A segment of Biolek window function for a state close to hard-switching mode; (b) time diagram of the memristance M.

2.1.4. A Memristor Model Based on the Boundary Conditions

The Boundary Condition Memristor (BCM) model, proposed by Corinto and Ascoli [11], could be expressed with the following system of two equations (2.16) [11]:

$$
\frac{dx}{dt} = k\,i\,f_{BCM}(x)
$$
$$
v = i\,[R_{ON}x + R_{OFF}(1 - x)]
\tag{2.16}
$$

where the first equation in System of Equations (2.16) is the state differential equation, f_{BCM} is the applied window function, and the second one is the state-dependent Ohm's Law of the memristor element. The BCM model could realistically represent the memristor boundary effects [11].

The window function for this model is a special case of the Biolek window function for very high values of the positive integer exponent. The applied window function f_{BCM} [11] is presented with the following voltage-dependent and state-dependent Equations (2.17):

$$f_{BCM}(x) = 1, \quad x \in (0, 1)$$
$$f_{BCM}(x) = 1, \quad x = 0 \;\&\; v \geq v_{thr}$$
$$f_{BCM}(x) = 0, \quad x = 0 \;\&\; v < v_{thr} \qquad (2.17)$$
$$f_{BCM}(x) = 1, \quad x = 1 \;\&\; v < -v_{thr}$$
$$f_{BCM}(x) = 0, \quad x = 1 \;\&\; v \geq -v_{thr}$$

The BCM model is suitable for investigating memristor elements both for soft-switching and hard-switching regimes [11]. For the soft-switching mode, the state variable x does not achieve its limiting values—zero and unity. For the hard-switching regime, the state variable x reaches its limiting values, and for a forward-biased memristor, if the state variable x is equal to 0, its assessment could be altered only if the direction of the electric current flowing through it and accordingly that of the voltage across the memristor element become positive [11].

If the applied voltage is negative, the boundary between the doped and the undoped regions of the memristor element could not move in the left direction due to physical and mechanical restrictions [4,11]. If the state variable x grows up and becomes unity, the border between the doped and the undoped regions of the memristor could not be moved in the right direction due to physical limitations [4,11]. The state variable x could be altered only if the electric current and the corresponding voltage across the memristor are negative. For a reverse-biased memristor element, operating in a hard-switching state, if the state variable x becomes zero, it could be altered if the electric current (and accordingly the voltage) is negative. If the state variable x becomes unity, it could be changed if the current flowing through the memristor is positive [11].

The time graphs of the applied sine-wave memristor voltage v, the corresponding current i flowing through the memristor, the flux linkage Ψ and the state variable x are presented in Figure 19(a,b) for further discussion. The current has only positive values, due to the rectification effect of the memristor element, operating in a hard-switching mode. The state variable x is a double-sided limited function and reaches its limiting values. The state–flux relationship, shown in Figure 20(a), is expressed by a multi-valued hysteretic loop. The curve describing the corresponding current–voltage characteristic is non-symmetrical (Figure 20(b)). In this case, the memristor behaves as a semiconductor diode and could be used for high-level and low-frequency signal rectification.

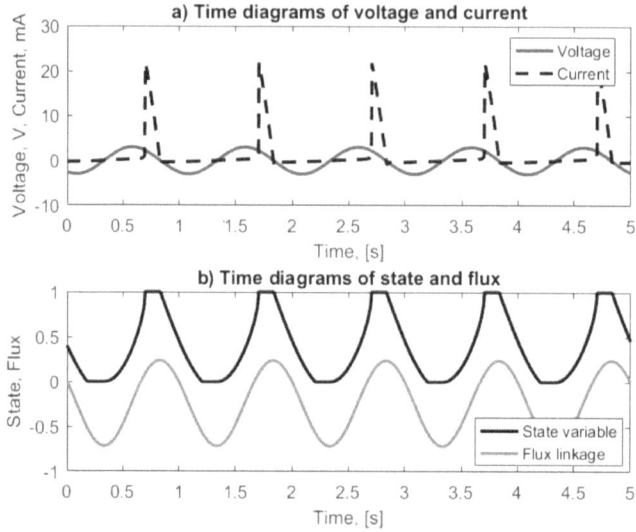

Figure 19. (a) Time diagrams of memristor voltage $v(t) = 3\sin\left(2\pi \times 0.5t - \frac{2\pi}{3}\right)$ and the current according to the BCM model with a boundary-condition-dependent rectangular window function; (b) time diagrams of the state variable x and the flux linkage Ψ where the state variable x reaches its limiting values, and the flux linkage Ψ has a sinusoidal shape.

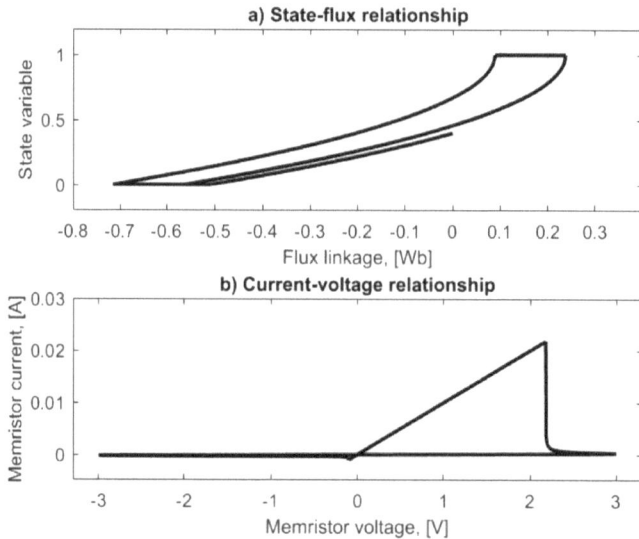

Figure 20. (a) State–flux characteristic of the memristor element according to the BCM model expressed by a multi-valued curve; (b) Current–voltage relationship of the memristor element, expressed by a non-symmetrical function; the memristor element has a rectification effect.

30

The window function for the BCM model is presented in Figure 21(a). It has a rectangular form and represents the boundary effects [11]. The time diagram of the memristance is presented in Figure 21(b). It is observable that the memristance of the element changes in the whole possible range from the ON-resistance to the OFF-resistance of the memristor element.

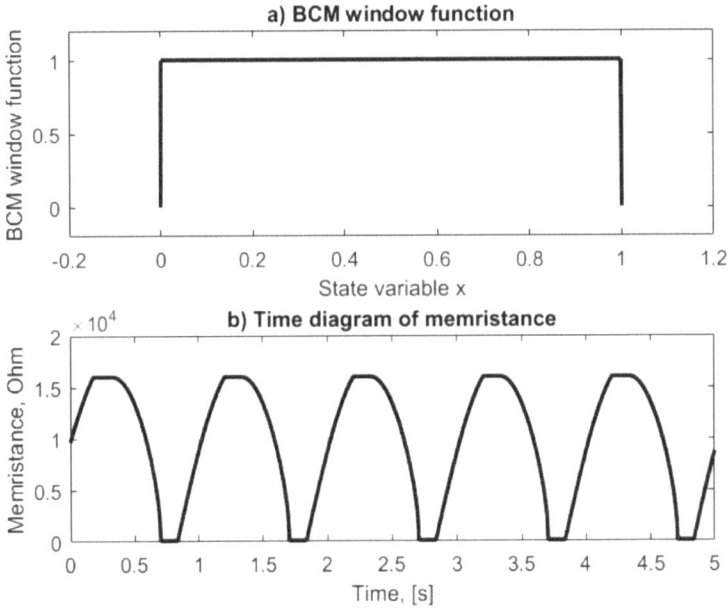

Figure 21. (a) BCM rectangular window function, derived for a hard-switching mode; (b) time diagram of the resistance of the memristor (memristance) according to the BCM model for a hard-switching mode. The resistance reaches its limiting values.

The memristor element is also investigated for the pulse voltage according to the BCM model [11]. The time graphs of the voltage and the corresponding state variable are shown in Figure 22. The state variable x is proportional to the time integral of the applied memristor voltage. It is observable that, when the voltage v is zero, the corresponding memristor state variable x retains its previous value. This result confirms the memory effect of the analyzed memristor element.

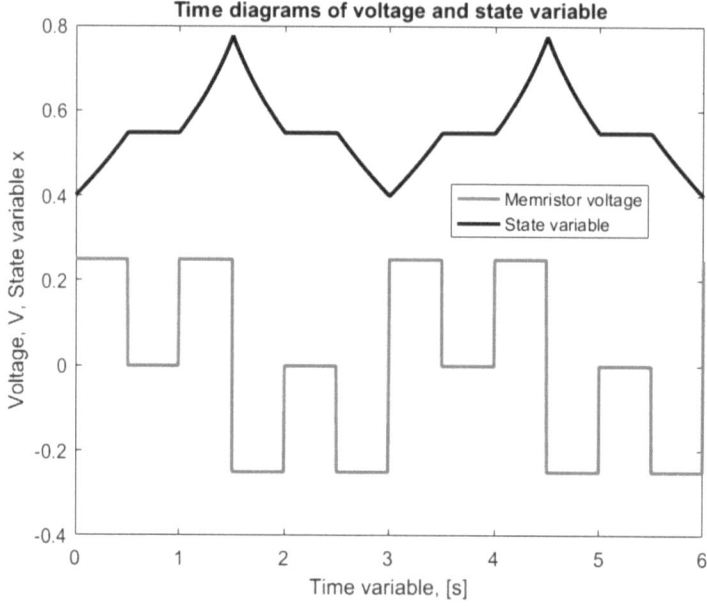

Figure 22. Time diagrams of the applied pulse voltage v and the corresponding state variable x; the state variable x is proportional to the time integral of memristor voltage v. the memristor holds its previous value of the state variable x when the voltage v is zero, which is the so-called memory effect.

2.1.5. Generalized Memristor Model Based on the Boundary Conditions

The Generalized Boundary Condition Memristor (GBCM) model [14] is highly similar to the previous one—the BCM model [11]. The main difference between them is only the use of activation threshold v_{thr} not only for the boundaries, but for all the values of the memristor state variable x in the GBCM model [14]. The corresponding window functions for the universal boundary condition model is described by System of Equations (2.17) [14].

The other equations describing the analyzed memristor element according to the GBCM model are the same as the ones used in Section 2.1.4 for the BCM model—System of Equations (2.17).

$$\begin{aligned}
f_{GBCM}(x) &= 1, &&\text{if } x > 0 \text{ and } x < 1 \text{ and } |v| \geq v_{thr}\\
f_{GBCM}(x) &= 0, &&\text{if } x > 0 \text{ and } x < 1 \text{ and } |v| < v_{thr}\\
f_{GBCM}(x) &= 1, &&\text{if } x = 0 \text{ and } v \geq v_{thr}\\
f_{GBCM}(x) &= 0, &&\text{if } x = 0 \text{ and } v < v_{thr}\\
f_{GBCM}(x) &= 1, &&\text{if } x = 1 \text{ and } v < -v_{thr}\\
f_{GBCM}(x) &= 0, &&\text{if } x = 1 \text{ and } v > -v_{thr}
\end{aligned} \tag{2.18}$$

2.1.6. Introducing a Polarity Coefficient in the Idealized Memristor Models

The memristor behavior depends on its biasing—forward or reverse connection. This property is especially important, if a given circuit contains more than one memristor elements [5,15,16]. An example for such a situation is presented in Figure 23—an anti-series memristor circuit with two memristor elements. The current flowing through the memristors has the same direction for both the first and the second elements. This current decreases the resistance of the first memristor, and at the same time, it increases the resistance of the second one. This fact confirms the behavior of the memristor as a polar element, which has a cathode and an anode electrodes. The memristors presented in Figure 23 are coupled in a series. The cathode of the first memristor is connected to the cathode of the second memristor. Due to this, the circuit is named "an anti-series biasing". The polarity of the memristors is important property both for soft-switching and hard-switching modes. Due to the described polarity of the memristor element, a modification has to be introduced to the describing System of Equations (2.1), which is shown as the following System of Equations (2.19):

$$\frac{dx}{dt} = \eta k i f(x)$$
$$v = i\left[R_{ON}x + R_{OFF}(1-x)\right]$$

(2.19)

where η is a polarity coefficient; for a forward-biased memristor, it has a value of 1, and for reverse-biasing it is equal to -1 [15,16]. System of Equations (2.19) describing the anti-series memristor circuit (Figure 23), after transformation, is as follows—System of Equations (2.20):

$$\begin{vmatrix} v_{M1} = \left[(R_{ON} - R_{OFF})x_1 + R_{OFF}\right]i, & \frac{dx_1}{dt} = ki \\ v_{M2} = \left[(R_{ON} - R_{OFF})x_2 + R_{OFF}\right]i, & \frac{dx_2}{dt} = -ki \end{vmatrix}$$

(2.20)

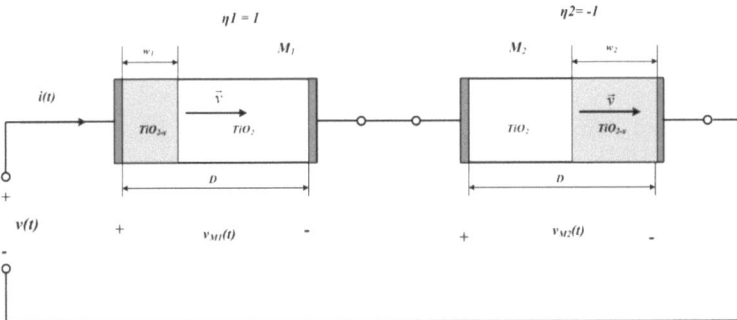

Figure 23. An anti-series circuit with two memristors. The circuit contains polar elements connected in reverse biasing, that is, the cathode of the first memristor is connected to the cathode of the second element.

After applying the KVL for the analyzed circuit given in Figure 23 ($v_{M1} + v_{M2} = v$) and System of Equation (2.20), the next System of Equations (2.21) is derived:

$$\left| \begin{array}{l} [(R_{ON} - R_{OFF})(x_1 + x_2) + 2R_{OFF}]dx_1 = k\,v(t)dt \\ -[(R_{ON} - R_{OFF})(x_1 + x_2) + 2R_{OFF}]dx_2 = k\,v(t)dt \end{array} \right. \qquad (2.21)$$

The primary values of the state variables of the memristors are x_{10} and x_{20}. Using the finite difference method [14,16] Equations (2.11) and (2.12) are solved numerically in MATrix LABoratory (MATLAB) environment [13]. Using System of Equations (2.21), the initial values of state variables x_{10} and x_{20} and Kirchhoff Voltage Law (KVL) [4,12,17] after transformations, the equivalent resistance of the anti-series memristor circuit could be theoretically acquired:

$$R_{eq} = \frac{v}{i} = (R_{ON} - R_{OFF})(x_{10} + x_{20}) + 2R_{OFF} = constant \qquad (2.22)$$

Since the equivalent resistance of the circuit R_{eq} is not dependent on the applied voltage v or the current i, the anti-series memristor circuit as a whole is a linear one. Let the initial values of the state variables be x_{10}' and x_{20}' and the equivalent resistance of the circuit is R_{eq}' at the moment $t = 0$. Due to the boundary effects, if $x_{10}' \neq x_{20}'$, it is sometimes possible for one of the memristors to reach and retain a fully open or fully closed state for a given time interval. After this interval, it is possible for the other memristor element to still operate in an active mode ($0 < x < 1$). If, at this moment, the source voltage changes its direction, then it could be assumed that a new process starts, but with different initial values of the state variables. The equivalent resistance of the circuit will reach a different value: $R_{eq}'' \neq R_{eq}'$. If both the memristors start to operate in a hard-switching mode, then the equivalent resistance reaches and retains its new stable and constant value: $R_{eq} = R_{ON} + R_{OFF}$. Only in the special case, when $x_{10}' = 1 - x_{20}'$, the equivalent resistance of the circuit has a stable and constant value of $R_{eq} = R_{ON} + R_{OFF}$ for the general electric mode. The diagrams of the state variables of the memristors for the hard-switching and soft-switching modes are shown in Figure 24(a,b).

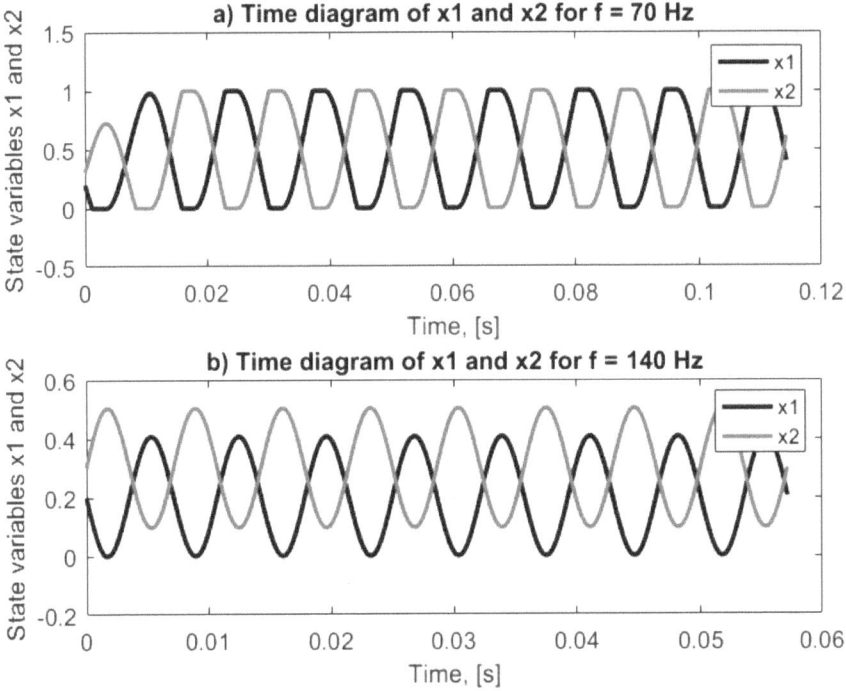

Figure 24. (a) Time diagrams of the state variables of the memristors M_1 and M_2 for the hard-switching mode with a magnitude of 4.3 V and a frequency of 70 Hz; **(b)** time diagrams of the state variables of the memristors M_1 and M_2 for the soft-switching mode with a magnitude of 4.3 V, and a frequency of 140 Hz. For anti-series connection, if the resistance of the first memristor decreases, at the same time, the memristance of the second element grows up.

The memristors are switched to the anti-phase mode. When the state variable of the first memristor increases, at the same time, the state variable of the second memristor diminishes, and vice versa. For the hard-switching mode, if the state variable of the first memristor has a value of zero, at the same time, the state variable of the second memristor has a value of unity. Sometimes, if the initial values of the state variables are chosen with specific values, an equivalent constant resistance of the anti-series circuit could be acquired. The corresponding time diagrams of the applied voltage, the current and the equivalent resistance of the circuit are presented in Figure 25(a,b), for both hard-switching and soft-switching regimes. The equivalent resistance acquires a constant value during its operation. The state–flux relationships of the single memristor elements for hard-switching and soft-switching states are given in Figure 26(a,b).

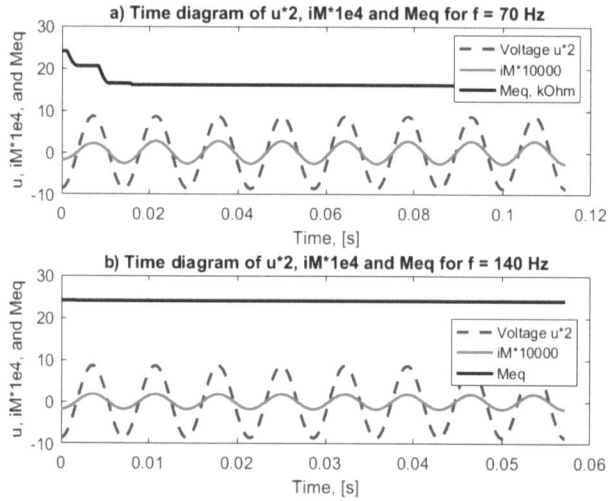

Figure 25. (**a**) Time graphs of the sine-wave voltage, applied to the memristor circuit with a magnitude of 4.3 V and a frequency of 70 Hz for the hard-switching mode, the current, and the equivalent resistance of the electric circuit, and (**b**) the time diagrams of the voltage, current, and the equivalent memristance of the memristor circuit, supplied by a sine voltage with s magnitude of 4.3 V and a frequency of 140 Hz; for the hard-switching mode, the equivalent resistance has a constant value; for the soft-switching mode, it is possible to obtain the equivalent resistance with a constant value if $x_1 + x_2 = 1$.

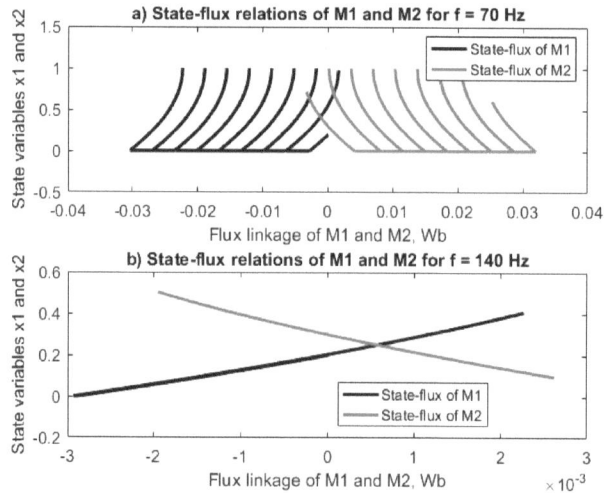

Figure 26. (**a**) State–flux relationships of the single memristor elements for hard-switching and (**b**) for soft-switching modes; due to the different biasing polarities of the memristors, the first state–flux relationship is expressed by an increasing curve, while the second one is illustrated by a decreasing curve.

Due to the different biasing polarities of the memristors, the first state–flux relation is shown by an increasing curve, while the second one is illustrated by a monotonically decreasing curve. For the hard-switching mode, the state–flux characteristic is a multi-valued function. The corresponding current–voltage relationships of the single memristors are presented in Figure 27(a,b). They both are pinched hysteresis loops.

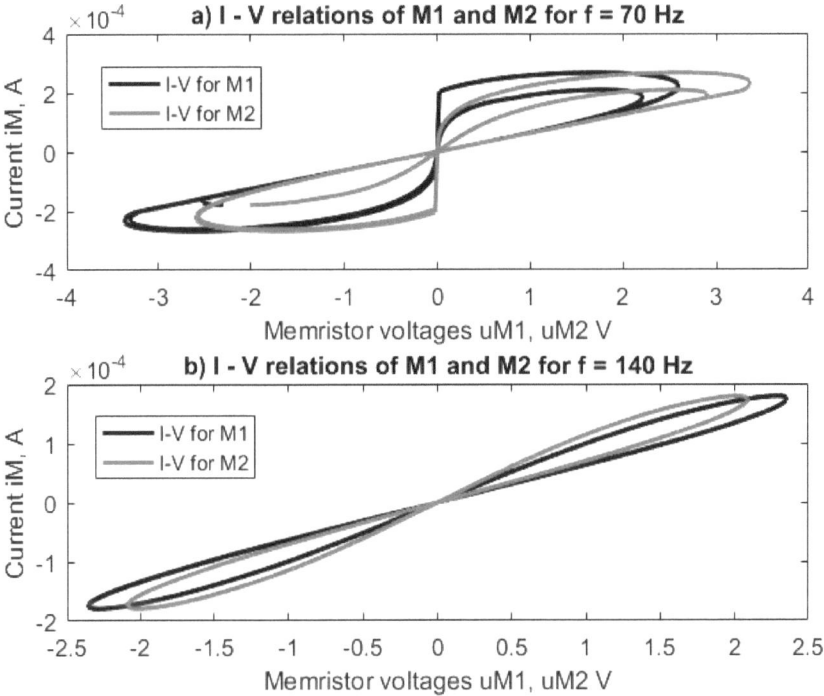

Figure 27. (a) Current–voltage relationships of the single memristor elements for hard-switching and (b) for soft-switching modes. The single memristors operate in a soft-switching mode and their current–voltage characteristics are pinched hysteresis loops, as shown in Figure 27(a,b). An interesting phenomenon is the behavior of the whole circuit in an electric field. At each instant time, the sum of the resistances of the memristors is a constant.

For the hard-switching mode, these characteristics are multi-valued functions, while for the soft-switching mode, they are single-valued graphs. The current–voltage relationship of the whole circuit for the soft-switching mode is illustrated by a straight line, while for the hard-switching mode, the corresponding current–voltage relationship is a double-valued function, as shown in Figure 28(a,b).

Due to this fact, the corresponding current–voltage relationship of the circuit is demonstrated by a straight line. This phenomenon could be explained with the deviations of the resistances of the memristors. The increase of the resistance of the first memristor is equal to the decrease of the resistance of the second memristor at each moment. For description of this specific circuit behavior, the simplified linear drift memristor model is especially appropriate [4].

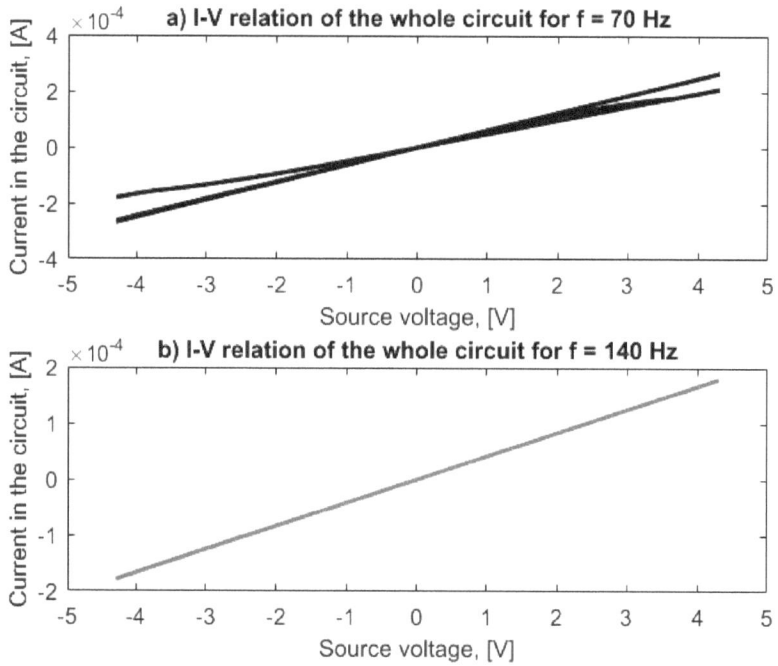

Figure 28. Current–voltage relationship of the memristor circuit, for hard-switching (a) and for soft-switching modes (b); for the soft-switching mode, the current–voltage characteristic is shown by a straight single-valued line, while for the hard-switching mode, it is a multi-valued function, due to the change of the equivalent resistance of the memristor circuit.

2.1.7. Pickett Memristor Model

The Pickett model is based on physical measurements and analyses of the mechanism of the electric current flowing through a thin tunnel barrier in insulating oxides [7]. At this point of view, the Pickett model is a physical nonlinear model [7,17]. It is a highly nonlinear memristor model. The structure of a memristor cell according to the Pickett [7,17] and Simmons models [18] is shown in Figure 29(a) for further discussion and clarification of the basic parameters and the memristor behavior in electric fields.

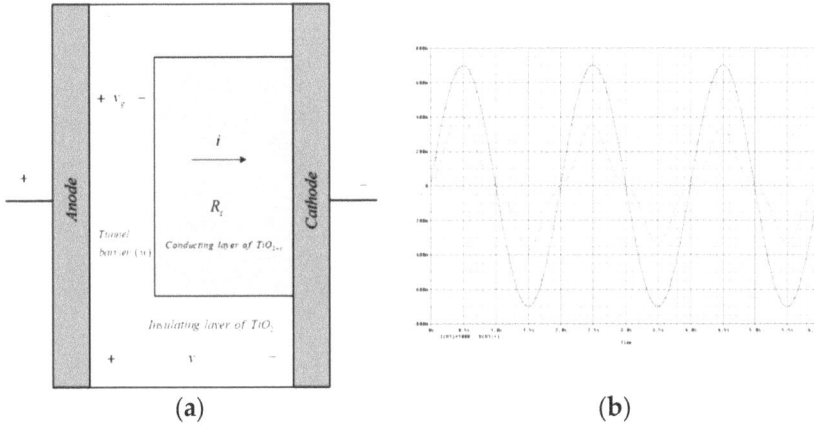

(a)   (b)

Figure 29. (a) A simplified memristor nanostructure according to Pickett and Simmons models, presenting the metallic electrodes, the formed conducting layer of doped titanium dioxide, the tunnel barrier and the insulating layer of titanium dioxide in the memristor volume; (b) time diagrams of the memristor voltage and current for the sinusoidal voltage signal with a frequency of 0.5 Hz and an amplitude of 0.68 V.

The memristor electrodes are made of titanium or platinum and the dielectric layer is completed by pure amorphous TiO_2 material. The conducting conduit in the memristor element is formed by a thin region of TiO_2 material saturated with oxygen vacancies, derived by the forming process [4,7]. There is also a thin tunnel wall with a length of w. The resistance of the conducting channel is approximately equal to $Rs = 215\ \Omega$ [7]. The Pickett memristor model is based on the Simmons model of the electrical current flowing through a thin tunnel wall in an insulating material [18]. According to the Simmons model, the memristor could operate mainly in two states—ON-state and OFF-state [7,18]. Time diagrams of the memristor voltage with a frequency of 0.5 Hz and an amplitude of 0.68 V and the current in sinusoidal mode are shown in Figure 29(b) for representation the nonlinear form of the memristor current i [19].

When the memristor is in the open state (OFF-resistance; $i > 0$), the differential equation of the state variable w has to be expressed by the next Equation (2.23) [7]:

$$\frac{dw}{dt} = f_{off}\sinh\left(\frac{i}{i_{off}}\right)\exp\left[-\exp\left(\frac{w - a_{off}}{w_c} - \frac{|i|}{b}\right) - \frac{w}{w_c}\right] \qquad (2.23)$$

39

The memristor parameters used in Equation (2.23) and their respective values [7] are given in Table 1. The corresponding numerical values of these parameters are experimentally derived [7,17,18].

Table 1. Parameters for a memristor element in the OFF-resistance state.

Quantity	f_{off}	i_{off}	a_{off}	B	w_c
Measurement unit	µm/s	µA	nm	µA	pm
Value	3.5	115	1.20	500	107

If the element is in the ON-state (ON-resistance; $i < 0$), the state differential equation of the state variable w is the following Equation (2.24) [7]:

$$\frac{dw}{dt} = f_{on}\sinh\left(\frac{i}{i_{on}}\right)\exp\left[-\exp\left(\frac{w - a_{on}}{w_c} - \frac{|i|}{b}\right) - \frac{w}{w_c}\right] \tag{2.24}$$

The corresponding characteristics and the numerical evaluations for the ON-state [7] are given in Table 2. These parameters of the analyzed memristor are established by laboratory experiments [7,17,18].

Table 2. Basic parameters for the ON-switched memristor element.

Quantity	f_{on}	i_{on}	a_{on}	B	w_c
Measure	µm/s	µA	nm	µA	pm
Value	40	8.9	1.80	500	107

The electric current through the tunnel wall of the memristor element [7,17,18] is presented in the next Equation (2.25):

$$i = \frac{j_0 A}{\Delta w^2}\left[\Phi_1 \exp\left(-B\sqrt{\Phi_1}\right) - (\Phi_1 + |v_g|)\exp\left(-B\sqrt{\Phi_1} + |v_g|\right)\right] \tag{2.25}$$

where v_g is the voltage drop across the tunnel wall [7]. This voltage drop could be evaluated using the Kirchhoff's Voltage Law (KVL) [12] and the voltage v across the memristor element [7] could be expressed by the next Equation (2.26):

$$v_g = v - iR_s \tag{2.26}$$

The quantity j_0 [7] could be illustrated with Equation (2.27):

$$j_0 = \frac{e}{2\pi h} = \frac{1.6 \times 10^{-19}}{2 \times 3.14 \times 6.63 \times 10^{-34}} = 3.84 \times 10^{13}, \ [C/(J \cdot s)] \tag{2.27}$$

where e is the electric charge of the electron, and h is the Planck's constant.

The surface area of the tunnel joint [7,17,18] is $A = 10^4$ nm^2. The variation of the length of the tunnel junction [7] is described by the next Equation (2.28):

$$\Delta w = w_2 - w_1 \tag{2.28}$$

The second term in Equation (2.28) [7] is presented in the following Equation (2.29):

$$w_1 = 1.2 \times \frac{\lambda w}{\Phi_0}, \; nm^2 \tag{2.29}$$

where $\Phi_0 = 0.95$ V is the height of the described potential wall. The variable λ [7] in Equation (2.29) is expressed by the next Equation (2.30):

$$\lambda = \frac{e \ln 2}{8 k \pi \varepsilon_0 w \times 10^{-9}}, \; V \tag{2.30}$$

where $k = 5$ is the permittivity of the TiO$_2$ medium [7,17], and ε_0 is the absolute permittivity of the vacuum space, which is equal to 8.85×10^{-12} F/m [12].

Substituting Equation (2.30) into Equation (2.29), it is derived that $w_1 = 0.126$ nm. According to Reference [7], the next Equation (2.31) is acquired by using Equation (2.30):

$$w_2 = w_1 + w \left(1 - \frac{9.2 \times \lambda}{4\lambda + 3\Phi_0 - 2|v_g|} \right), \; nm \tag{2.31}$$

The measure B [7] is expressed with the following Equation (2.32):

$$B = \frac{4\pi \sqrt{2me} \, \Delta w \times 10^{-9}}{h}, \; V^{-\frac{1}{2}} \tag{2.32}$$

where m is the electron weight [7]. The variable Φ_1 [7,18] is presented with the following Equation (2.33):

$$\Phi_1 = \Phi_0 - |v_g| \frac{w_2 + w_1}{w} - \frac{1.15 \times \lambda w}{\Delta w} \ln \left[\frac{w_2(w_1 - w)}{w_1(w_2 - w)} \right], \; V \tag{2.33}$$

The Pickett memristor model [7,17,18] is fully described with Equations (2.23)–(2.33). The memristor circuit under test contains a sinusoidal voltage source and a memristor element. The sinusoidal voltage signal produced by the voltage source and applied for the present investigation is: $v(t) = 0.6 \sin(2\pi t)$.

The acquired flux–charge relationship rendering to the Pickett memristor model [17,19] is shown in Figure 30 for observation of one of the main memristor characteristics. In the present case of analysis, the flux–charge relationship is shown by a multi-valued curve [17,19]. The corresponding current–voltage characteristic is given in Figure 31. It is also a multi-valued pinched hysteresis curve. Based on the present investigations, it is established that for voltage signals higher than 0.75 V, many convergence issues appear [19], although the Pickett model has the highest accuracy and sometimes it is used as a reference memristor model [7,17].

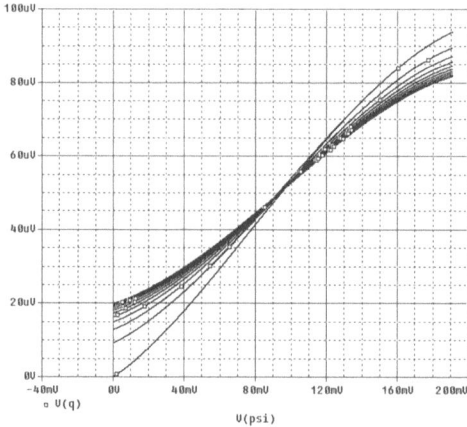

Figure 30. Flux–charge characteristic of the element according to the Pickett memristor model, derived for the memristor voltage $v(t) = 0.6\sin(2\pi t)$. It is one of the main characteristics of the memristor, and in this case, it is a multi-valued curve.

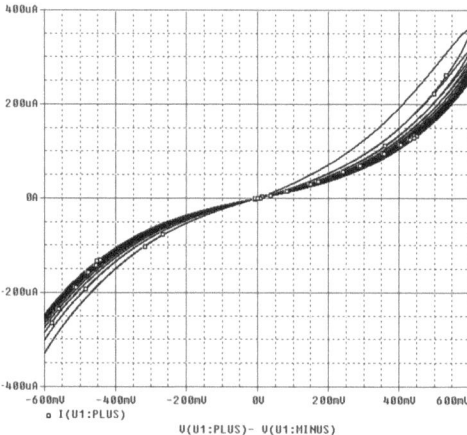

Figure 31. Current–voltage characteristic according to Pickett memristor model, presenting the behavior of the memristor for the sinusoidal voltage $v(t) = 0.6\sin(2\pi t)$. In the present analysis, it was derived as a multi-valued curve.

2.1.8. Other Nonlinear Memristor Models

More appropriate memristor device models without convergence problems have been proposed [17,18,20,21]. In References [20,21], a physical memristor model is described. It is based on experimental results [20]. The approximated relationship between the memristor current i and the applied memristor voltage v is presented by the next nonlinear Equation (2.34) [20,21]:

$$i = x^n \beta \sinh(\alpha v) + \chi[\exp(\gamma v) - 1] \qquad (2.34)$$

where $\alpha = 2\ \text{V}^{-1}$, $\beta = 0.9\ \mu\text{A}$, $\gamma = 0.0004\ \text{V}^{-1}$ and $\chi = 1.10^{-4}\ \mu\text{A}$ are fitting parameters, and $n = 1$ is a factor that determines the influence of the state variable x on the electric current. In this discussed memristor model [20–22], the state variable x is a normalized parameter in the interval [0, 1]. This memristor model presents an asymmetric performance [20].

When the memristor element is in the ON-state, the state variable is near unity and the electric current is dominated by the first term in Equation (2.34). It illustrates a tunneling performance [17,18]. If the memristor element is in the OFF-status, the state variable x is near zero and the current is primarily represented by the second term in Equation (2.34), which describes a diode equation [20,21].

The applied memristor model [20,21] uses a nonlinear function for the voltage in the corresponding state differential Equation (2.35) [17,20,21]:

$$\frac{dx}{dt} = a\, f(x)\, v^m \qquad (2.35)$$

where $a = 1\ \text{V}^{-m}{\cdot}\text{s}^{-1}$ is a constant, $m = 9$ is an odd integer exponent, and $f(x)$ is a window function applied for approximate description of the nonlinear ionic motion and the boundary effects [6,11]. The window function introduces nonlinearity with respect to the state variable x of the memristor element [11]. Equations (2.34) and (2.35) completely define the corresponding physics-based memristor model [20,21].

The ion transport in the memristor element is linked to the ionic dopant drift in the corresponding material [22]. The ions of the oxygen vacancies jump between two neighboring positions via a migration barrier [23,24]. This potential barrier could be decreased by the applied electric field [20,21,23]. The ions can acquire more thermal energy by Joule heating and they can therefore straightforwardly overcome the tunnel barrier [23,24]. The nonlinearity of the ionic dopant drift starts from local Joule heating or high electric fields [23,25]. The used window function gives an approximate relationship between the state change and the electric current [6,23]. The present nonlinear model was investigated for the sinusoidal memristor voltage $v(t) = 1.5\sin\left(2\pi \times 2t - \frac{2\pi}{3}\right)$. The derived state–flux relationship is presented in Figure 32. It represents a multi-valued hysteresis loop, due to the boundary effects and the applied window function [6]. The corresponding current–voltage relationship is presented in Figure 33 for expression of the memristor non-symmetric behavior. In this case, the memristor operates in hard-switching mode [11].

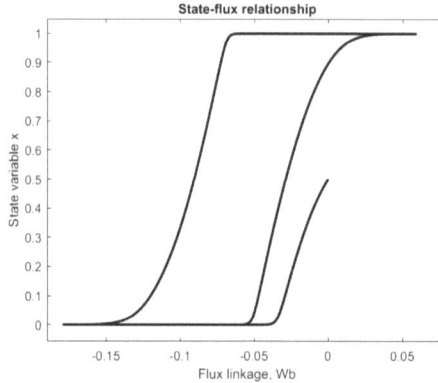

Figure 32. State–flux relationship of the memristor according to the analyzed model together with a Biolek window; the applied signal is: $v(t) = 1.5 \sin\left(2\pi \times 2t - \frac{2\pi}{3}\right)$, the state-flux relation is described by a multi-valued hysteresis curve and represents a hard-switching operation.

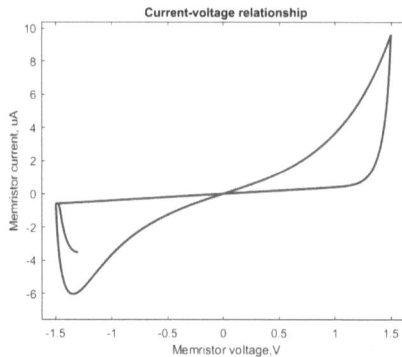

Figure 33. The corresponding current–voltage characteristic of the described nonlinear model, representing asymmetric memristor behavior for the hard-switching state. The applied memristor voltage is described as: $v(t) = 1.5 \sin\left(2\pi \times 2t - \frac{2\pi}{3}\right)$.

It was established by several analyses that, for very low voltages, the corresponding current–voltage relationship is a non-hysteretic nonlinear function. The described nonlinear memristor model is based on physical measurements and it is capable to realistically represent the phenomenon of exponential ionic drift. In the real memristor elements, increasing the voltage with several hundred mV leads to increase of the switching rate by several orders of magnitude [24]. Several classic memristor models are not able to illustrate this phenomenon [21,24].

2.2. Author's Modified Memristor Models

2.2.1. Dependence between Charge Mobility and Temperature of the Titanium Dioxide Memristor Element

In Reference [26,27], an investigational diagram of the dependence between oxygen vacancies mobility and the absolute temperature T of a pure TiO_2 material is offered. It is exciting that in this oxide material, the slope of this experimentally derived function is positive. The ionic drift mobility grows up with the increased temperature [26–28]. The oxygen vacancies mobility at room temperature is $\mu_v = 1.10^{-14}$ (m^2/V·s). By the use of these physical data, an interpolation polynomial with the lowest degree of -18 is acquired in MATLAB [13]. The approximated function is presented by Equation (2.36):

$$\mu_v = 3 \times 10^{-12} + 1 \times 10^{29} \left(-0.0005 \times T^{-18} - 0.2394 \times T^{-17} \right) \tag{2.36}$$

The electric charge mobility grows up with the increased memristor temperature. The relationship between the oxygen vacancies mobility and the temperature measured in degrees Celsius is presented in Figure 34(a) and is used for further description. The operation of the memristors, especially the switching rate, depends on ambient temperature [27,28].

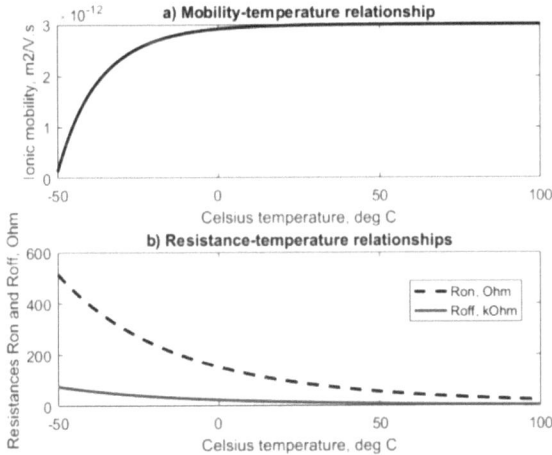

Figure 34. (a) Dependence between oxygen vacancies mobility in the amorphous titanium dioxide material and the temperature measured in Celsius; **(b)** dependence between the memristor' resistances in the open state (R_{OFF}) and in the closed state (R_{ON}) and the temperature measured in Celsius t.

2.2.2. Relationship between the Resistances of the Memristor Element in OFF- and ON-States and the Temperature Measured in Celsius

An investigational relationship between the specific conductivity of titanium dioxide σ and the absolute temperature T is shown in Reference [28]. The variables on the coordinate axes are in logarithmic degree [26,28]. The relationship between them is approximately represented by Equation (2.37) [26–28]:

$$\lg \sigma = 5.72 \lg T - 9 \tag{2.37}$$

where σ is the specific conductivity of the amorphous titanium dioxide material. Using Equation (2.36) and the expression of the specific resistances (2.37), the resistances R_{OFF} and R_{ON} (Equations (2.38) and (2.39), respectively) are acquired [27,28]:

$$R_{OFF} = 2.06 \times 10^{18} T^{-5.72} \tag{2.38}$$

$$R_{ON} = 6.33 \times 10^{16} T^{-6} \tag{2.39}$$

In Figure 34(b), the relationships between R_{OFF} and R_{ON} and the temperature t are presented. It is obvious that the resistances of the memristor R_{OFF} and R_{ON} decrease lightly with the increased memristor temperature. The resistance in the closed state diminishes with the increased memristor temperature. The characteristics of these relationships are related to the thermal generation of charge carriers in this type of semiconductor material. The change of these resistances is non-desirable in an operation mode [26–28].

2.2.3. Investigation of the Internal Diffusion Processes in the Strukov and Williams Titanium Dioxide Memristor Element

This concentration gradient of the oxygen vacancies causes diffusion processes from the saturated layer to the undoped region of the memristor element, which is equivalent to flow of diffusion current [27,29,30]. The density of this diffusion current (unit: A/m^2) is presented by the Fick's first law [29]—Equation (2.40):

$$J_{V_O} = -qD\frac{\partial N}{\partial x} \tag{2.40}$$

where q is the charge of an oxygen vacancy and can be written as: $q = 2e = 3.2 \times 10^{-19}$ C, D (unit: m^2/s) is the diffusion coefficient of the dopant ions, N (unit: m^{-3}) is the volumetric concentration of oxygen vacancies in the memristor element, and x is the coordinate in which location the diffusion is realized [27,28]. The diffusion coefficient of the oxygen vacancies is presented by Equation (2.41) [26,28,30]:

$$D = \frac{\mu_V k_B T}{q} \tag{2.41}$$

where k_B is the Boltzmann constant and it is equal to 1.3787×10^{-23} J/K. The dependence between the diffusion coefficient D and the temperature t is drawn in MATLAB [13] using Equations (2.36) and (2.41) is presented in Figure 35. It can be used if the memristor operates at different ambient temperatures. It is obvious that the diffusion coefficient D grows up quickly with the increased temperature. This phenomenon could be physically explained by the increase of the kinetic energy of the oxygen vacancies and their simplified penetration from the doped layer to the depleted (undoped) region of the memristor element.

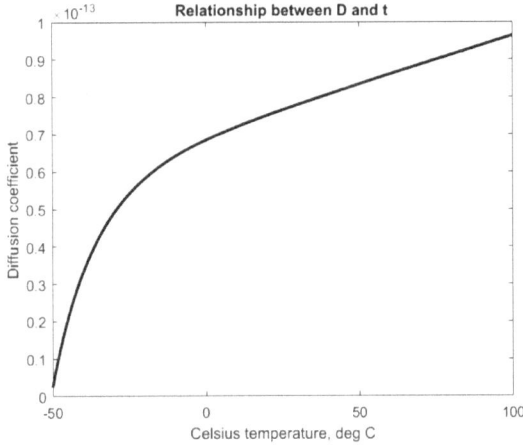

Figure 35. Dependence between the diffusion coefficient of the oxygen vacancies in the titanium dioxide material D and the temperature measured in Celsius t of the memristor one-port element.

The volume of a memristor element could be calculated as a volume of rectangular parallelepiped with a length l of 10 nm, a width a of 50 nm and a height b of 50 nm by the following Equation (2.42):

$$V_{total} = l \times a \times b = 10 \times 10^{-9} \times \left(50 \times 10^{-9}\right)^2 = 2.5 \times 10^{-23} \ [m^3] \qquad (2.42)$$

The volume of the doped region of the memristor element is described by the next Equation (2.43):

$$V_{doped} = l_{doped} \times a \times b = 0.1 \times 10 \times 10^{-9} \times \left(50 \times 10^{-9}\right)^2 = 2.5 \times 10^{-24} \ [m^3] \ (2.43)$$

The density of titanium dioxide material [29] is ρ = 4230 kg/m³. Its molar mass is $M(TiO_2)$ = 79.86 [g/mol] [29]. The mass of a molecule of TiO_2 is expressed by the next Equation (2.44):

$$m_{TiO_2} = \frac{M(TiO_2)}{N_A} = \frac{79.88}{6.02 \times 10^{23}} = 1.327 \times 10^{-22} \ [g/molecule] \qquad (2.44)$$

where N_A is the Avogadro constant and it is equal to 6.02×10^{23} mol^{-1} [29]. The number of the TiO$_2$ molecules in a volume of 1 m^3 is given in the following Equation (2.45):

$$N_{TiO_2} = \frac{\rho(TiO_2)}{m_{TiO_2}} = \frac{4230}{1.33 \times 10^{-25}} = 3.19 \times 10^{28} \ [m^{-3}] \tag{2.45}$$

The volumetric concentration of the oxygen vacancies in the doped region of the memristor element is presented in the next Equation (2.46) [29]:

$$N_{V(O)} = z \times N_{TiO_2} = 0.03 \times 3.19 \times 10^{28} = 9.56 \times 10^{26} \ [m^{-3}] \tag{2.46}$$

where z is the stoichiometric coefficient in the material TiO$_{2-z}$. The number of oxygen vacancies in the saturated layer of the element is given in the next Equation (2.47):

$$n_{V(O)} = V_{doped} N_{V(o)} = 2.5 \times 10^{-24} \times 9.56 \times 10^{26} = 2390 \tag{2.47}$$

The ionic diameter d of an oxygen vacancy has an approximate value of 0,2 nm [28,29]. The length of the saturated region w is 1 nm. The number of the atomic layers in the doped region is expressed by the following Equation (2.48):

$$n_{at.layers} = \frac{w}{d} = \frac{1}{0.2} = 5 \tag{2.48}$$

Then, the number of oxygen vacancies closest to the border between the two sub-regions of the element is presented in the next Equation (2.49):

$$n_{V(at.layer)} = \frac{n_{V(O)}}{n_{at.layers}} = \frac{2390}{5} = 478 \tag{2.49}$$

The surface concentration of oxygen vacancies Q_0 close to the border between the doped and the undoped layers of the memristor element is expressed by Equation (2.50) [27,29]:

$$Q_0 = \frac{n_{V(at.layer)}}{a \times b} = \frac{478}{(50 \times 10^{-9})^2} = 1.91 \times 10^7 \ [m^{-2}] \tag{2.50}$$

The Fick's second law [29,30] for the process of diffusion in the memristor is presented in Equation (2.51):

$$\frac{\partial N_{V(O)}}{\partial t} = D \frac{\partial^2 N_{V(O)}}{\partial x^2} \tag{2.51}$$

In the case presented here, a diffusion process from a limited source of dopant ions exists. The initial conditions (dopant concentration) according to the dopant (charge) concentration are expressed by Equation (2.52):

$$N(t, x) = N_S, \ t = 0, \ x = 0 \tag{2.52}$$

In the initial moment and for the left boundary of the memristor, the oxygen vacancies concentration has a maximal value of N_S. The respective limiting conditions (dopant concentration) are described by Equation (2.53):

$$N(t, x) = 0, \ t > 0, \ x \to \infty \tag{2.53}$$

The time derivative of the charge concentration near to the left boundary is expressed by Equation (2.54):

$$\frac{dN}{dx} = 0, \ t \in (0, \infty), \ x = 0 \tag{2.54}$$

The result derived after solving Equation (2.51) at the initial and the limiting conditions presented above is given in the next Equation (2.55) [27,29,30]:

$$N(x, t) = N_S^I \exp\left(-\frac{x^2}{4Dt}\right) [m^{-3}] \tag{2.55}$$

where N_S^I is the instant volumetric concentration of oxygen vacancies near the border between the saturated and undoped regions of the memristor [29,30] and it is written in the next Equation (2.56) as:

$$N_S^I = \frac{Q_O}{\sqrt{\pi Dt}} [m^{-3}] \tag{2.56}$$

With the increased diffusion time, the ion's concentration in the depleted layer decreases. This is due to the penetration of the dopant in the volume of the memristor and the depletion of oxygen vacancies. With the increased distance from the dopant source, the concentration of oxygen vacancies decreases, but after a very long time, the concentration of the dopant ions equalizes in the whole volume of the memristor cell. With growing up of the memristor temperature, the charge carriers' mobility increases, but the conductances of the memristor at open and closed states increase. As an outcome, at an absolute temperature T of 400 K, the whole charge that could be accumulated by the memristor element is about 100 times lower than that at room temperature. The electric current through the investigated memristor is higher than the electric current at room temperature (about 20 degrees Celsius). Therefore, the switching speed of the element grows up as well and the operational state changes.

The properties and the characteristics of the memristor element as an electronic switch are worse at high ambient temperature.

The diffusion processes in the Strukov Williams' titanium dioxide memristor are parasitic phenomena and exist due to the concentration gradient around the boundary between the doped and undoped regions [27]. The coefficient of diffusion D increases with the increased ambient temperature. This is an undesired process because the internal diffusion in the memristor causes losses of stored information. For example, if a logical zero has been written in the memristor cell (in this case, the resistance of the element is close to R_{OFF}), after a very long time interval, due to the diffusion of oxygen vacancies, the dopant ions penetrate from the doped region in the whole volume of the memristor, so that the resistance of the memristance will acquire a new value—R_{ON}. The described process of losing information initially stored in the memristor is quicker at very high temperatures. In the end, it could be concluded that the increasing of the ambient and operating temperatures of Williams memristor has a negative influence on its parameters. For improvement of the memristor's characteristics in an operation mode, the application of a cooling device is recommended [27].

2.2.4. Analysis of Memristor's Parasitic Parameters and Mutual Inductances between Neighboring Elements of a Memristor Crossbar

The equivalent circuit of two neighboring memristors placed on a crossbar is presented in Figure 36 for clarification of the parasitic currents and explanation of the influence of the parasitic phenomena on the normal operation of the memristor elements. The parasitic capacitances C_1 and C_2 represent these two memristors' own capacitances, respectively, due to the overlapping between the memristors' electrodes. The parasitic inductances L_1 and L_2 present the inductances of each of the electrodes of the memristor crossbar, respectively [31]. The coefficient of mutual inductance between the memristor elements M is almost equal to each of the own inductances due to the full embracement of the magnetic flux by both the platinum rims of the memristor elements. The coefficient of magnetic influence k has a value very close to unity. In this investigation, three values of the coefficient k are used: $k = 0.90$, $k = 0.95$ and $k = 0.99$ [12,31].

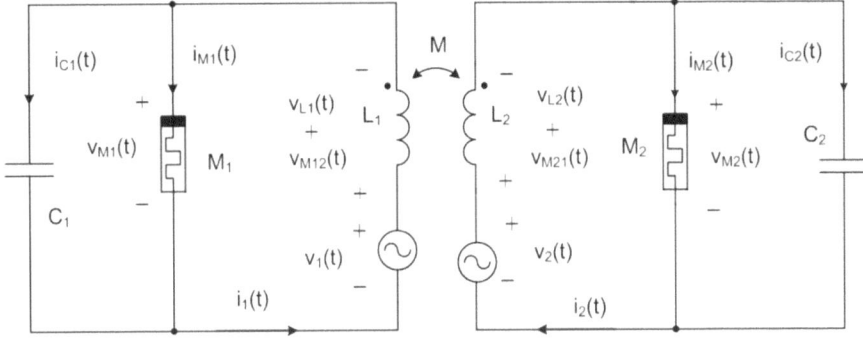

Figure 36. An equivalent circuit of two neighboring memristors with a memory crossbar for clarification of the appearance of parasitic capacitances and inductances and their influence on the normal operation of the considered memristor crossbar.

The capacitances C_1 and C_2 are calculated as a capacitance of a flat capacitor. The parasitic capacitance is a combination of two capacitances of the doped and the un-doped regions connected in a series. The values of the relative dielectric permittivities of the doped and the un-doped regions of the memristor, (ε_{r1} and ε_{r2}), respectively, are 170 and 150 [4,12]. The width of the memristor electrodes a is 50 nm. The lengths of the doped (D_1) and undoped (D_2) regions of the memristor are $D_1 = w_1 = 1\ nm$, and $D_2 = D - w_1 = 9\ nm$, respectively. The equivalent capacitance between the memristor electrodes is given by Equation (2.57) [12]:

$$C_{par} = \frac{C_1 C_2}{C_1 + C_2} = \frac{\varepsilon_0 \varepsilon_{r1} \frac{a^2}{D_1} \varepsilon_0 \varepsilon_{r2} \frac{a^2}{D_2}}{\varepsilon_0 \varepsilon_{r1} \frac{a^2}{D_1} + \varepsilon_0 \varepsilon_{r2} \frac{a^2}{D_2}} \tag{2.57}$$

The numerical result for the parasitic capacitance of the memristor is $C_{par} = 3.10^{-16}\ F$. The parasitic own inductance is calculated by solving a definite double-sided integral with the use of electromagnetic field theory. A conductor with a finite length l is disposed over the z-axis. A permanent current with intensity i flows through the wire. A current element idl is placed in the center of a Cartesian coordinate system (Figure 37) [12,31]. The induction lines of the magnetic field are concentric circles which are placed in planes, parallel to the coordinate x0y plane.

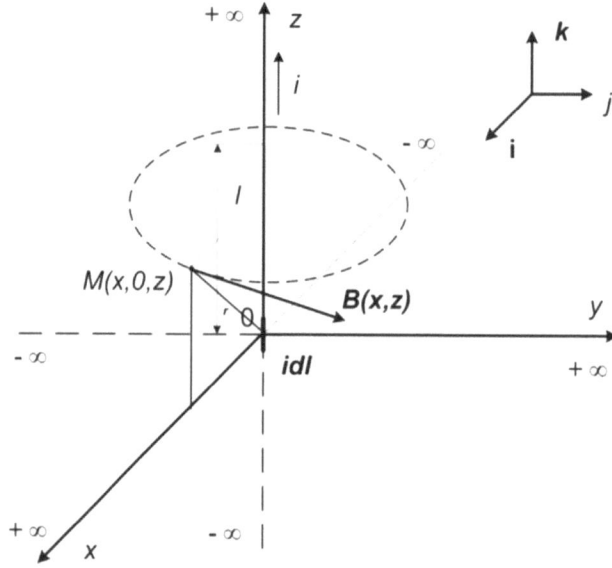

Figure 37. Three-dimensional coordinate geometry for deriving the coefficient of own inductance of the Williams titanium dioxide memristor and mutual inductance between the considered memristor elements.

The tangent vector to the circle $B(x,z)$ at a given point is the vector of magnetic flux density. It is determined by the next Equation (2.58)—the Biot-Savart Law [12,31]:

$$d \overrightarrow{B}(r) = \frac{\mu_0\, i\, dl}{4\,\pi\, r^2} \overrightarrow{k} \times \overrightarrow{e}_r \tag{2.58}$$

where $\overrightarrow{B}$ is the magnetic flux density vector, μ_0 is the magnetic permeability of vacuum medium, r is the radius-vector of the previously chosen point M, k is the unity vector parallel to the z-axis, and e_r is the unity vector, parallel to the present radiusvector of the point M. The radius of the sphere on which the chosen point M lies has a length given by Equation (2.59) [12,31]:

$$r = \sqrt{x^2 + z^2} \tag{2.59}$$

The single vector on the radius-vector of point M is presented by Equation (2.60) with geometrical considerations [12,31]:

$$\overrightarrow{e}_r = \frac{x\, \overrightarrow{i} + z\, \overrightarrow{k}}{\sqrt{x^2 + z^2}} \tag{2.60}$$

After substituting Equations (2.59) and (2.60) into Equation (2.58), the following Equation (2.61) for the elementary magnetic flux density is derived [12,31]:

$$d\,\vec{B}(x,z) = \vec{j}\,\frac{\mu_0\,i}{4\,\pi}\,\frac{x}{(x^2+z^2)^{\frac{3}{2}}}\,dz \tag{2.61}$$

where $\vec{j}$ is the unity vector parallel to the y axis. The full magnetic flux acquired by the current element idl is presented by Equation (2.62), using definite double-sided integration [31]:

$$d\,\Phi = \int\limits_{-\infty}^{+\infty}\int\limits_{0}^{+\infty} d\vec{B}(x,z)\,dx\,dz\,\vec{j} = \int\limits_{-\infty}^{+\infty}\int\limits_{0}^{+\infty} \vec{j}\,\frac{\mu_0\,i}{4\pi}\,\frac{x}{(x^2+z^2)^{\frac{3}{2}}}\,dz\,dx\,dz\,\vec{j} =$$
$$= dz \int\limits_{-\infty}^{+\infty}\int\limits_{0}^{+\infty} \frac{\mu_0\,i}{4\pi}\,\frac{x}{(x^2+z^2)^{\frac{3}{2}}}\,dx\,dz \tag{2.62}$$

The full magnetic flux generated by the wire with a limited length of l is acquired after integration with respect to the z coordinate and it is expressed by Equation (2.63):

$$|\Phi| = \frac{\mu_0 i}{4}\int\limits_{0}^{l} dz = \frac{\mu_0 i}{4}l \tag{2.63}$$

The own inductance of the memristor element is given by the next Equation (2.64), taking into account the lengths of the respective memristor electrodes [31]:

$$L = \left|\frac{\Phi}{i}\right| = \left|\frac{\frac{\mu_0 i}{4}l}{i}\right| = \frac{\mu_0}{4}l = \frac{4\pi\times10^{-7}}{4}\times 0,03 = \tag{2.64}$$
$$9,4248 \times 10^{-9}\,H = 9,4\,[nH]$$

The parasitic inductance of the memristor element in the center of the analyzed memory crossbar is $L = 9.4$ nH. The mutual inductance between the memristors M is calculated with involvement of the mutual magnetic flux between the corresponding parallel memristor wires. Its value is close to the numerical value of the memristor's own inductance. The parameters of the second memristor were chosen, of which the values are 5% larger than those of the first memristor element parameters, and they are expressed by Equation (2.65):

$$C_2 = 1.05 \times C_1 = 1.05 \times 3 \times 10^{-16} = 3.15 \times 10^{-16} \text{ F}$$
$$L_2 = 1.05 \times L_1 = 1.05 \times 9.4 \times 10^{-9} = 9.87 \text{ nH}$$
$$v_{m1} = v_{m2} = 1 \text{ V} \qquad (2.65)$$
$$R_{ON2} = 1.05 \times R_{ON1} = 1.05 \times 100 = 105 \ \Omega$$
$$R_{OFF2} = 1.05 \times R_{OFF1} = 1.05 \times 16 = 16.8 \text{ k}\Omega$$

The mutual inductance between two neighboring memristors is calculated with the use of the coefficient of magnetic connection k and the own inductances of the memristors [12,31] and it is expressed by Equation (2.66):

$$M_1 = k_1 \sqrt{L_1 L_2} = 0.90 \times \sqrt{9.4 \times 10^{-9} \times 9.87 \times 10^{-9}} = 8.67 \ nH$$

$$M_2 = k_2 \sqrt{L_1 L_2} = 0.95 \times \sqrt{9.4 \times 10^{-9} \times 9.87 \times 10^{-9}} = 9.15 \ nH \qquad (2.66)$$

$$M_3 = k_3 \sqrt{L_1 L_2} = 0.99 \times \sqrt{9.4 \times 10^{-9} \times 9.87 \times 10^{-9}} = 9.54 \ nH$$

The parasitic parameters were found to have very low values. After several analyses, it was established that the parasitic parameters—own and mutual inductances and capacitances—do not strongly affect the normal operation of the memristor crossbars. Only for very high frequencies, these parameters have to be taken into account [31].

2.2.5. A Nonlinear Memristor Model with a Sensitivity Threshold and a Changeable Window Function

The new memristor model proposed in this paper is based on both Joglekar and GBCM models. The model applies the Joglekar window function with a changeable positive integer exponent. The new idea in the proposed model is that the positive exponent of the Joglekar window function and the corresponding nonlinearity extent of the memristor model could be altered in the operation process of the proposed memristor in accordance to the applied electric field intensity [32].

The state differential equation for the memristor element according to Reference [5] is expressed by the next Equation (2.67):

$$\frac{dx}{dt} = k i f_J(x) = k i \left[1 - (2x - 1)^{2p}\right] \qquad (2.67)$$

where x is the state variable, k is a constant dependent only on the physical parameters of the memristor element, $f_J(x)$ is the Joglekar window function, and p is a positive integer exponent [5]. In the state differential equation shown above, the Joglekar window function is used for representation of the nonlinear ionic dopant drift of the memristor for high-intensity electric fields [5].

If the value of the integer exponent p increases, the window function tends to become almost parallel to the abscissa in the plateau region [5,32]. For very high values of the positive integer exponent p, the window function has an almost constant value of 1, except in the boundaries, where it has a value of zero. The approximate expression of the state-dependent current–voltage relationship for the memristor if $R_{OFF} \gg R_{ON}$ is presented in the next Equation (2.68) [4,5]:

$$v(t) = \left[R_{on}x + R_{off}(1 - x) \right] i(t) \approx R_{off}(1 - x)i(t) \tag{2.68}$$

where R_{OFF} and R_{ON} are the memristances for ON- and OFF-states, respectively [5]. After substituting Equation (2.68) into Equation (2.67), the current $i(t)$ from the state-dependent Ohm's Law as a function of the voltage signal $v(t)$ and the memristor state variable x, substituting the result in the state differential equation, and separating the variables x and t, the following expression (Equation 2.69) is derived [5,32]:

$$\frac{1 - x}{1 - (2x - 1)^{2p}}dx = \frac{k}{R_{OFF}}v(t)dt \tag{2.69}$$

Equation (2.66) above is numerically solved with the finite difference method in MATLAB [13]. The time diagrams of the memristor voltage—a pseudo-sinusoidal signal with exponentially increasing amplitude: $v(t) = 0.04e^t \sin(2\pi \times 1.7t - \frac{\pi}{4})$ and the corresponding memristor state variable x are given in Figure 38(a,b) for description of the memristor model behavior for different modes [32].

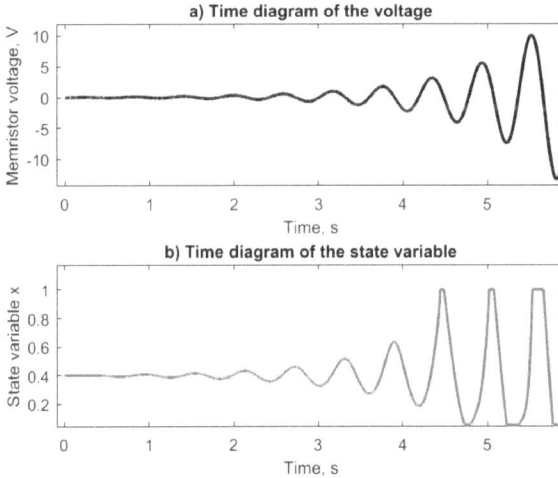

Figure 38. Time graphs of (**a**) the memristor voltage and (**b**) the state variable for signal $v(t) = 0.04e^t \sin(2\pi \times 1.7t - \frac{\pi}{4})$ for representation of the behavior of the suggested memristor model for a pseudo-sinusoidal voltage with an exponentially increasing magnitude; the soft-switching and hard-switching modes could be visually observed when the state variable becomes zero and unity, respectively.

If the voltage signal is lower than the activation threshold of the memristor, the corresponding state variable does not change in the time domain and the memristor has a constant resistance.

When the voltage becomes higher than the sensitivity threshold, then the state variable starts to change and the element operates in a soft-switching mode. In this situation, the state variable does not reach its limiting values. When the memristor voltage has very high amplitude, the element operates in a hard-switching state. In this case, the state variable reaches its limiting values. The corresponding curve of the memristor state variable is double-sided limited [32].

The state–flux and current–voltage characteristics acquired for the same voltage signal - $v(t) = 0.04e^t \sin(2\pi \times 1.7t - \frac{\pi}{4})$, as two of the basic memristor characteristics, are shown in Figure 39(a,b). It is obvious that for voltage values lower than the activation threshold of the memristor v_{thr} with a value of 40 mV, the state variable x remains unchanged and the memristor behaves as a linear resistor.

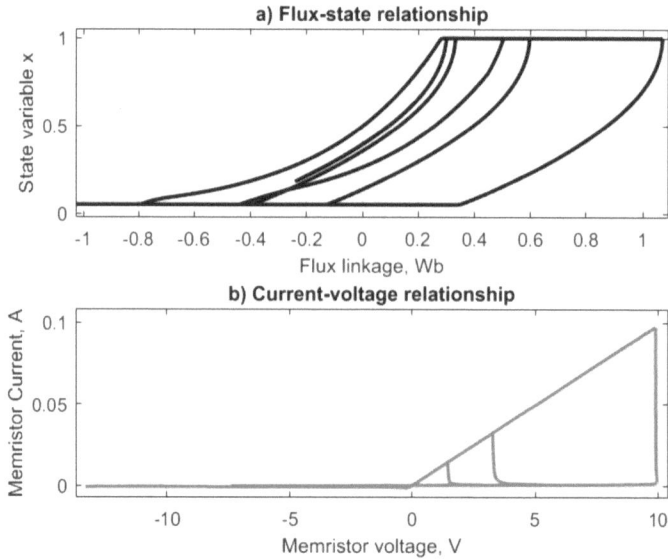

a) Flux-state relationship

b) Current-voltage relationship

Figure 39. (a) State–flux and (b) current–voltage characteristics of the memristor element for a pseudo-sinusoidal voltage signal with an exponentially-increasing amplitude, described as $v(t) = 0.04e^t \sin(2\pi \times 1.7t - \frac{\pi}{4})$. The derived characteristics express the transition between the soft-switching mode and the hard-switching mode when the voltage signal is lower than the memristor sensitivity threshold.

When the voltage exceeds the sensitivity threshold of the element v_{thr}, the memristor two-terminal component starts to operate in a soft-switching mode. For very high voltages, the memristor element operates in a hard-switching mode.

Then the state–flux relationship of the memristor is a multi-valued hysteresis curve and the corresponding current–voltage relationship is shown by an asymmetric curve with rectifying properties [32].

It was established by additional simulations that the transition between soft-switching and hard-switching regimes is associated with a rapid increase of the effective value of the memristor current.

2.2.6. A Model with Nonlinear Dopant Drift, a Modified Biolek Window Function and a Sensitivity Threshold

The basic suggestion of the proposed research is offering a new adapted memristor model with a strongly nonlinear dopant drift, appropriate for analysis of TiO_2 memristors for a large range of applied voltages [33]. For this reason, a mixture of the standard Biolek window function and a weighted sine-wave component is used [33]. The proposed altered model is based both on the GBCM model [14] and on the standard Biolek model [6]. The new modified memristor model proposed by the author has an improved assets—an increased extent of nonlinearity of the ionic motion due to the use of the additional normalized sinusoidal window component. The modified model is compared to the reference Pickett model, which is applied as a standard model. After the comparison, the altered Biolek memristor model is tuned with respect to Pickett model, and its main characteristics are almost identical to these of Pickett model [33]. The maximum amount of charges q_{max} that the element could accumulate is computed [12,33], when the border between the saturated and the undoped layers of the memristor is in the right edge of the whole memristor nanostructure and the state variable x has a value of unity [12,33], and is shown in the following Equation (2.70):

$$q_{max} = \int_0^\tau i(t)dt = \frac{1}{k}\int_0^1 dx = \frac{1}{k} = \frac{D^2}{\mu R_{ON}} = \frac{(10 \cdot 10^{-9})^2}{1 \cdot 10^{-14} \cdot 100} = 1 \cdot 10^{-4}\,C \qquad (2.70)$$

where the time for complete charging of the memristor element τ depends on the electric current i. It could be easily acquired that the charge memorized in the memristor element, given in Equation (2.71) is comparative to the maximum amount of charge q_{max} and the instantaneous value of the state variable x [12,33]:

$$q(x) = \int_0^{\tau_1} i(t)\,dt = \frac{1}{k}\int_0^x dy = q_{max}\,x = 10^{-4} \cdot x,\ C \qquad (2.71)$$

where τ_1 is the time for charging the memristor, which is less than τ; in the right side of Equation (2.69), the state variable x is substituted with the variable y, and x

is the upper limit of the applied definite integral [12]. The standard Biolek window function is shown in the next Equation (2.72) [6]:

$$f(x) = f_B(x) = -(x-1)^{2p} + 1, \quad v(t) \le 0, \quad [i(t) \le 0]$$
$$f(x) = f_B(x) = -x^{2p} + 1, \quad\quad v(t) > 0, \quad [i(t) > 0]$$

(2.72)

Here, a simple modification of Biolek window function is realized. To grow up the nonlinearity of the altered Biolek memristor model, the author proprosed an extra weighted sine-wave window function of the state variable of the memristor x, given in the Equation (2.73) [33]:

$$f(x) = f_{BM}(x) = \left[\frac{-(x-1)^{2p} + 1 + m\left(\sin^2(\pi x)\right)}{m+1} \right], \quad v(t) \le -v_{thr}$$

$$f(x) = f_{BM}(x) = \left[\frac{-x^{2p} + 1 + m\left(\sin^2(\pi x)\right)}{m+1} \right], \quad\quad v(t) > v_{thr}$$

(2.73)

$$f(x) = f_{BM}(x) = 0, \quad\quad\quad -v_{thr} \le v(t) \le v_{thr}$$

where $m = 0.2$ is a weight coefficient in front of the additional sinusoidal component of the window function, chosen after comparison of the results to experimental data [20,21]; it is established practically that for $m \in [0, 0.7]$, the behavior of the proposed memristor model is similar to those of the original Biolek model; if $m > 0.7$, and if the analysis is made for the hard-switching mode, the derived current–voltage relationship is illustrated by a multi-pinched hysteresis loop, and for higher values of m, the window function is similar to the Joglekar window, when it represents the soft-switching mode; v_{thr} is the activation threshold of the memristor and is set to 0.2 V, according to experimental measurements [20,21]; and f_{BM} is the modified Biolek window function of the memristor model [33]. The modified Biolek model proposed by the author is completely described with the following System of Equations (2.74) [33]:

$$\frac{dx}{dt} = \eta\, k\, i \left[\frac{-(x-1)^{2p} + 1 + m\left(\sin^2(\pi x)\right)}{m+1} \right], \quad v(t) \le -v_{thr}$$

$$\frac{dx}{dt} = \eta\, k\, i \left[\frac{-x^{2p} + 1 + m\left(\sin^2(\pi x)\right)}{m+1} \right], \quad\quad v(t) > v_{thr}$$

(2.74)

$$\frac{dx}{dt} = 0, \quad\quad\quad -v_{thr} \le v(t) \le v_{thr}$$

$$v = R\,i = [R_{ON}x + (1-x)R_{OFF}]\,i$$

where the first three equations of formula (2.74) represent the normalized modified window functions. The fourth equation in System of Equations (2.74) is the state-dependent Ohm's Law. Several analyses were made in MATLAB [13], using the numerical solution to System of Equations (2.74) by the finite difference method. After adjusting the model, the main results were acquired. The flux–charge relationship presented in Figure 40 is almost identical to that of the Pickett memristor model [7,33].

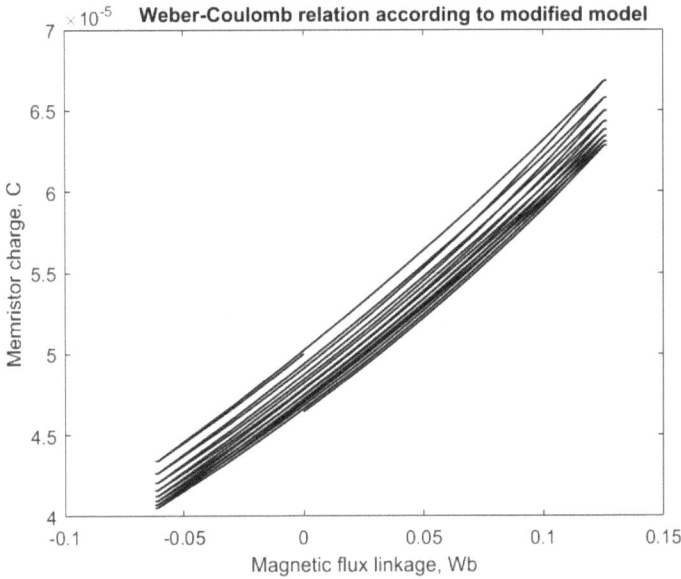

Figure 40. Flux–charge relationship according to the altered Biolek model. The applied memristor voltage is: $v(t) = 0.6\sin(2\pi t - \frac{\pi}{3})$; for the soft-switching mode, the acquired flux–charge characteristic of the element is shown by a multi-valued curve.

The working point of the memristor element in the field of the corresponding coordinate system moves among two segments of the window function in accordance to the direction of the applied voltage [33]. The corresponding current–voltage characteristic is shown in Figure 41 and is almost similar to that of the Pickett model, acquired in almost the same environment. The tuned values of the model coefficients are $p = 7$ and $m = 0.2$. The corresponding graphs of the applied voltage and state variable are given in Figure 42 for examination of the altering of the memristor state. The state variable x does not attain its limiting values and it could be accomplished that the memristor works in a soft-switching mode. The range of the state variable x is between 0.4 and 0.7.

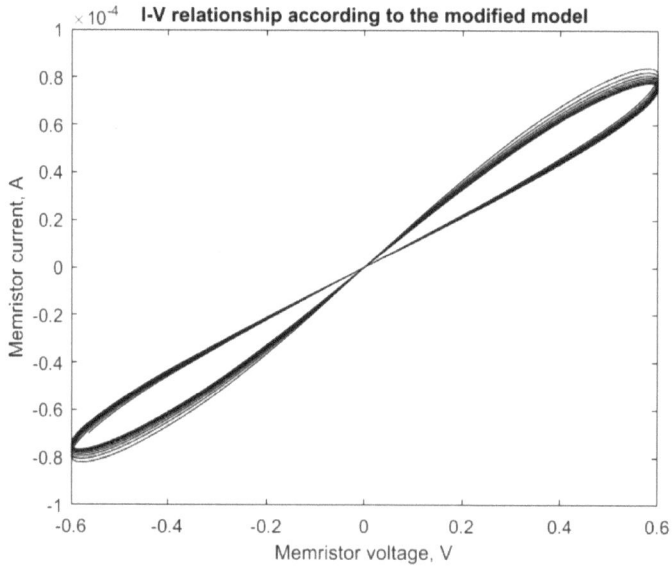

Figure 41. Current–voltage relationship of the memristor according to the adapted Biolek model. The applied memristor voltage is: $v(t) = 0.6 \sin\left(2\pi t - \frac{\pi}{3}\right)$; in this case, the memristor element is in a soft-switching regime.

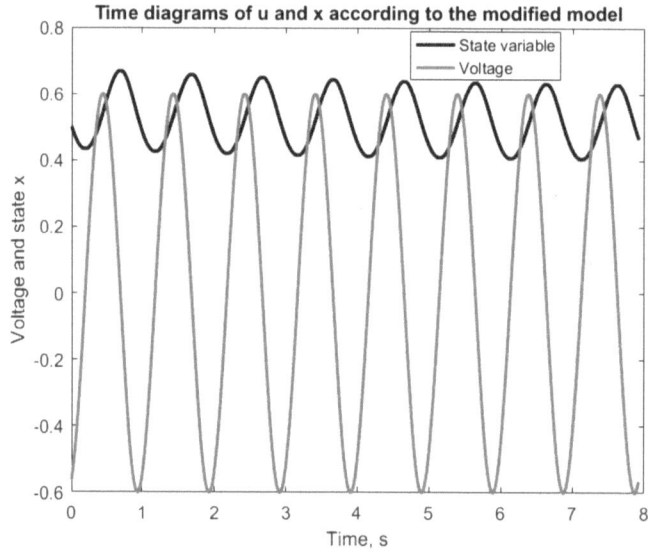

Figure 42. Time graphs of memristor voltage signal and state variable, shown for visual observation of the change of the state variable x for the soft-switching mode. The applied memristor voltage is: $v(t) = 0.6 \sin\left(2\pi t - \frac{\pi}{3}\right)$.

The dependence between the altered window function $f_{M(x)}$ and the state variable x is shown in Figure 43 [33].

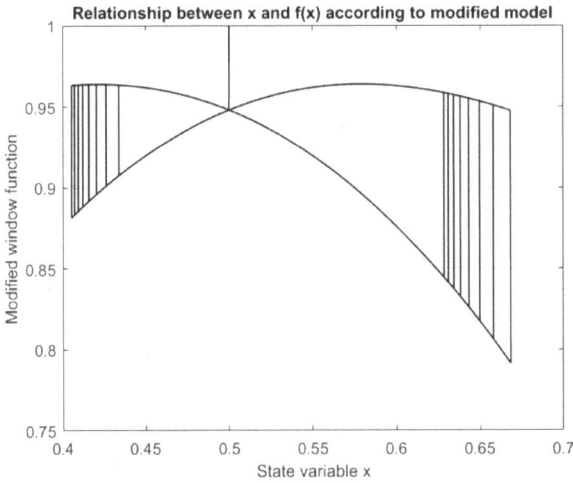

Relationship between x and f(x) according to modified model

Figure 43. The dependence between the improved window function and the state variable of the memristor element according to the altered Biolek model. The applied memristor voltage is: $v(t) = 0.6 \sin\left(2\pi t - \frac{\pi}{3}\right)$.

The tuned improved model was tested for a signal with a higher magnitude and with a negative initial voltage phase, that is, $v(t) = 3.6 \sin\left(2\pi t - \frac{2\pi}{3}\right)$. The corresponding flux–charge characteristic is given in Figure 44 for illustration of the hard-switching memristor regime. In this situation, it is illustrated by a multi-valued hysteresis loop. It attains its upper limit, corresponding to the memorized charge. The lower limiting value of the electric charge is almost attained. Then, it could be established that the memristor element operates in a state close to a hard-switching regime [33]. The corresponding current–voltage characteristic of the memristor element is shown in Figure 45 for presenting the memristor behavior in the field of the coordinates of the voltage v and the electric current i. In this case, the resistance of the memristor element for the applied positive voltage is comparatively low. The resistance of the memristor for the applied negative voltages is very high [33]. It could be concluded that if the element operates in a hard-switching state, then it behaves like a rectifier element. The time graphs of the memristor voltage and state variable for the hard-switching mode are shown in Figure 46 for visual observation of the shape of the state variable x. In this case, the state variable x attains its limiting values—0 and 1. The corresponding dependence between the altered window function $f_{M(x)}$ and the state variable x is presented in Figure 47. In this situation, the state variable values of the window function are in the interval between 0 and 1.

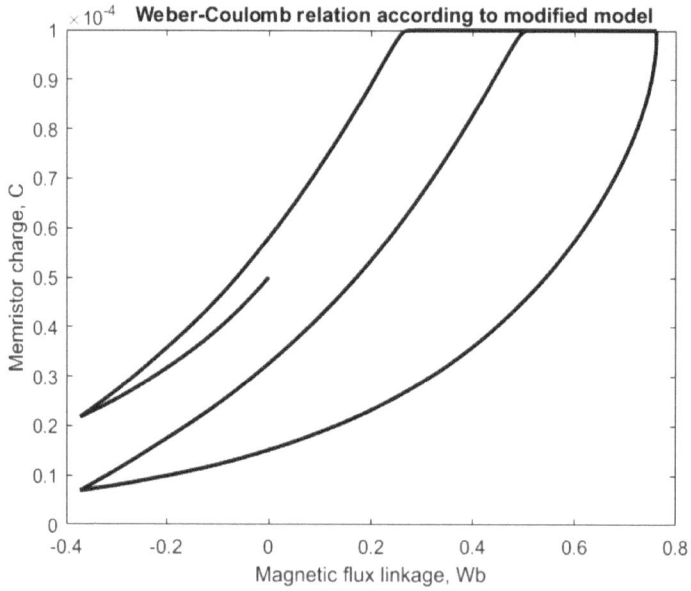

Figure 44. Weber–Coulomb characteristic according to the modified model for the hard-switching state. The applied memristor voltage is: $v(t) = 3.6 \sin\left(2\pi t - \frac{2\pi}{3}\right)$.

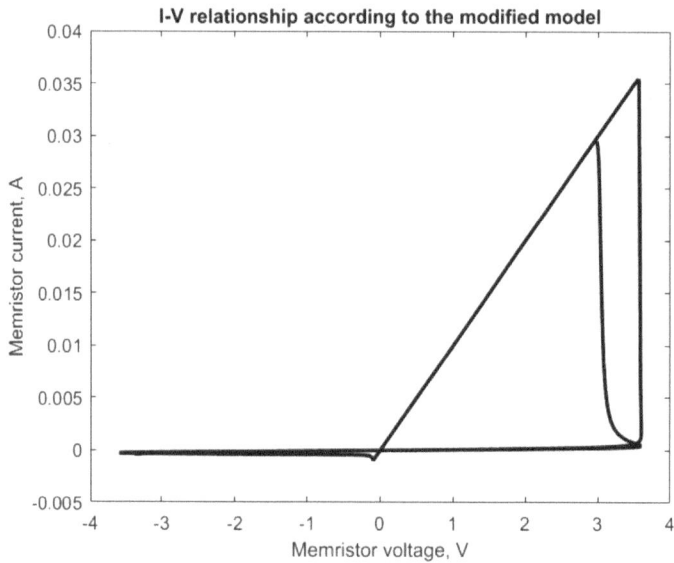

Figure 45. Current–voltage characteristic of the memristor element according to the altered model for the hard-switching regime. The applied memristor voltage is: $v(t) = 3.6 \sin\left(2\pi t - \frac{2\pi}{3}\right)$.

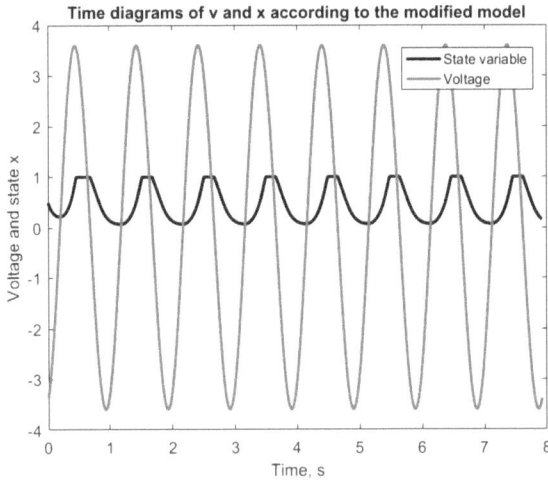

Figure 46. Time graphs of memristor voltage signal v and state variable x according to the modified model for the hard-switching regime. The applied memristor voltage is: $v(t) = 3.6\sin\left(2\pi t - \frac{2\pi}{3}\right)$.

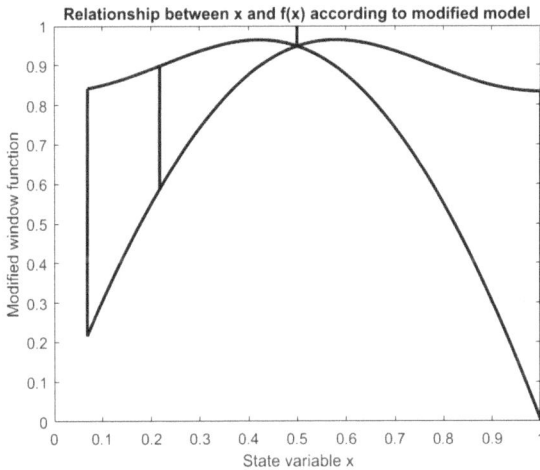

Figure 47. The modified window function for the hard-switching state. The applied memristor voltage is: $v(t) = 3.6\sin\left(2\pi t - \frac{2\pi}{3}\right)$.

The operating point of the memristor element in the field of the corresponding coordinate system moves on all the length of the two segments in accordance to the memristor voltage polarity [11,33].

The investigation of the improved Biolek memristor model for a voltage signal with an exponentially growing magnitude is conducted for confirmation of the expected performance of the memristor model with voltages lower and higher than the sensitivity threshold v_{thr}. The corresponding flux–charge characteristic in this case is shown in Figure 48. In the beginning, the working point of the memristor element in the field of the corresponding coordinate system moves on a straight line, because the state variable does not alter with voltages lower than the activation threshold. When the voltage signal is higher than the sensitivity threshold of the memristor, the operating point starts to move in a different segment of the flux–charge characteristic, which is a nonlinear function. The value of state variable x of the memristor element is altered in accordance to the flux linkage Ψ [11,33].

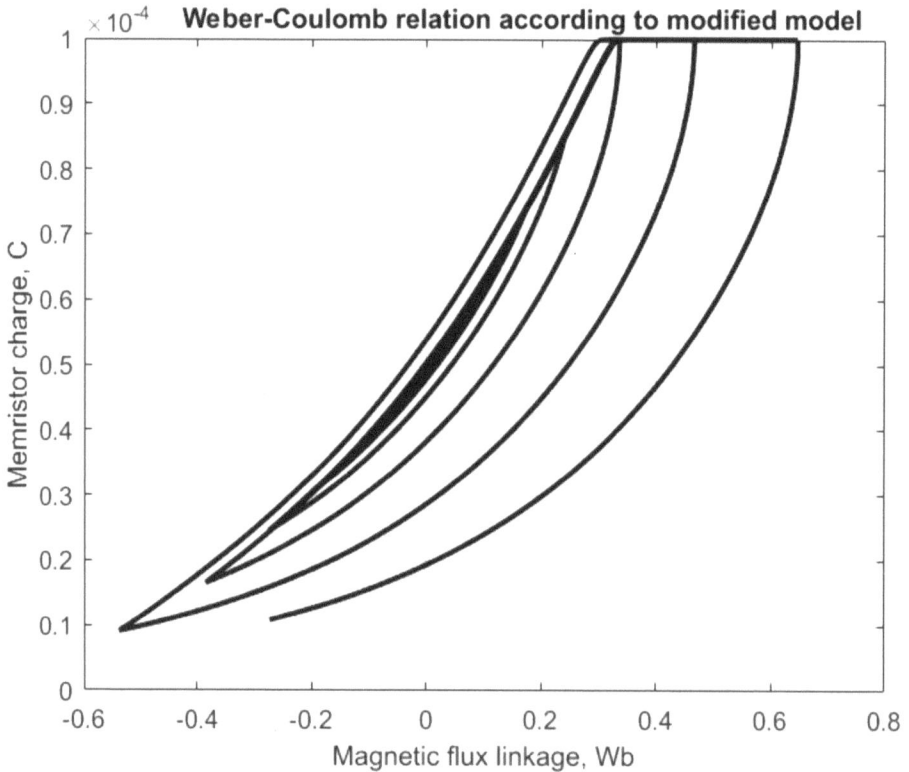

Figure 48. Flux–charge characteristic of the element according to the modified model for a pseudo-sinusoidal voltage with an exponentially growing magnitude, represented by $v(t) = 0.1e^t \sin(2\pi \times 3t)$.

The corresponding current–voltage characteristic is presented in Figure 49. In the beginning, it is shown by a straight line, and for voltages higher than the activation threshold, the curve representing the current–voltage relationship is a multi-valued pinched hysteresis loop. The corresponding time graphs of the memristor voltage and the state variable are given in Figure 50 for illustration of the transition between the different memristor states. The voltage signal is pseudo-sinusoidal with an exponentially growing magnitude. In the beginning, for voltages lower than the activation threshold, the state variable x does not change, but for higher voltages, it starts to change [33]. For voltages lower than the sensitivity threshold, the memristor element behaves like a linear resistor. If the voltage signal becomes higher than the sensitivity threshold, then the state variable x changes and the memristor element behaves like a standard memristor. The dependence between the improved window function $f_{M(x)}$ and the state variable x is shown in Figure 51. According to the voltage direction, the operating point of the memristor element in the field of the corresponding coordinate system moves between the two segments of the characteristics. After finishing the tuning process of the altered Biolek model and deriving the basic results, several conclusions could be made [33]. The improved model based on both the Biolek memristor model and GBCM model is a general one. It contains an altered window function, which is the sum of the standard Biolek window function, and a weighted sine-wave window function element. In the special case when the coefficient of the sine-wave window m and the sensitivity threshold v_{thr} are zero, the standard Biolek model is obtained.

If the improved memristor model shown here is tuned, we could derive results identical to those derived by the Pickett memristor model. Of course, the results derived in the present investigation are not precisely the same as those produced by the Pickett memristor model, which have the highest correctness but also have many convergence issues and is not suitable for computer simulations.

The basic benefit of the considered model according to the Pickett model is the absence of computational issues and the opportunity for its application in analysis of memristors and memristor circuits and devices [33]. The improved memristor model proposed by the author was compared to the Joglekar model for the soft-switching regime. The state–flux characteristic derived by the Joglekar model is shown by a single-valued graph, which is a benefit of the Joglekar model with respect to the altered memristor model which has multi-valued state–flux characteristics.

On the other hand, the altered model suggested in this research has several benefits, which the Joglekar model does not possess, that is, higher nonlinearity of the ionic drift, capability for realistic representation of the border effects, and the use of activation voltage threshold [33].

After comparison of the modified model to the Pickett model, it could be concluded that the latter could be simulated for a narrower voltage range. For voltages higher than 0.75 V, many convergence problems occur, and only operation in a soft-switching mode can be observed. An advantage of the modified model with respect to the Pickett memristor model is the possibility for simulation in a broad voltage range and the observation of the results for the hard-switching mode [33].

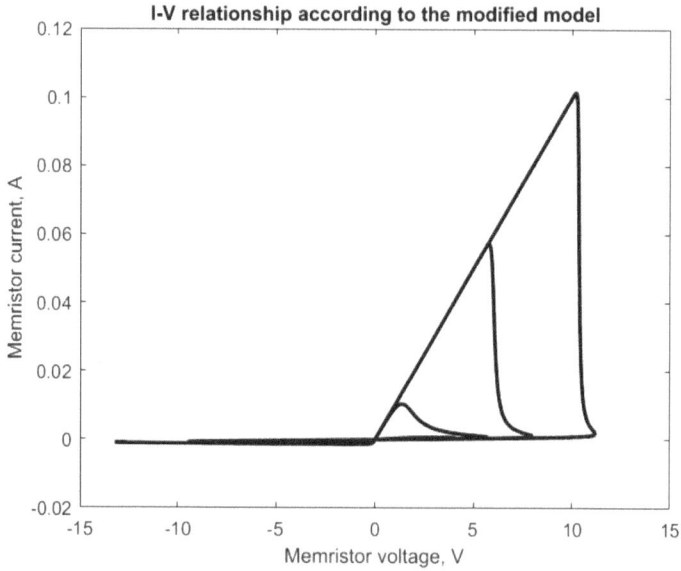

Figure 49. Current–voltage relationship according to the altered model for a pseudo-sinusoidal voltage with an exponentially increasing magnitude, expressed by $v(t) = 0.1e^t \sin(2\pi \times 3t)$.

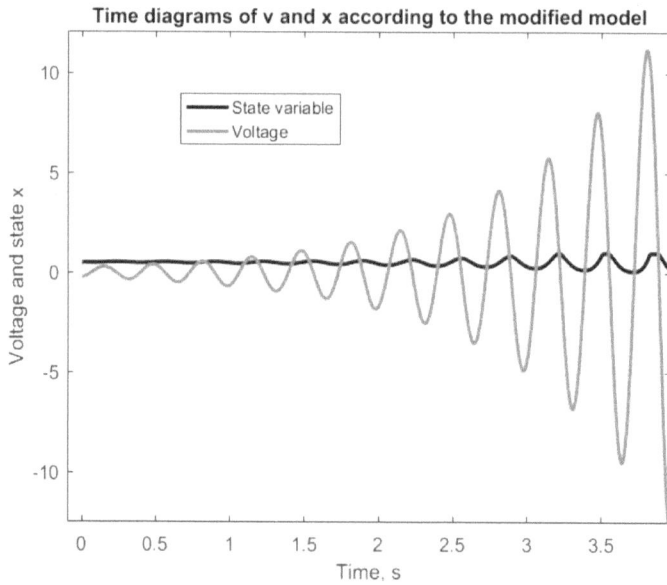

Figure 50. Time diagrams of the memristor voltage signal v and the state variable x according to the improved memristor model for a voltage with an exponentially increasing magnitude expressed by $v(t) = 0.1e^t \sin(2\pi \times 3t)$.

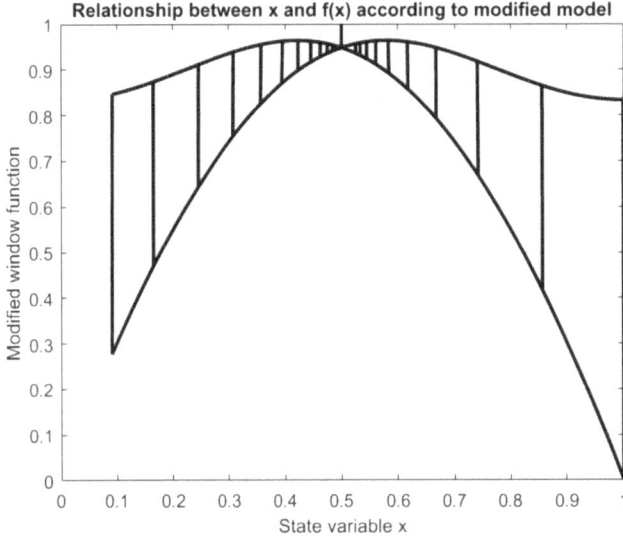

Figure 51. The modified window function $f_{M(x)}$ corresponding to a pseudo-sinusoidal memristor voltage with an exponentially increasing amplitude, expressed by $v(t) = 0.1e^t \sin(2\pi \times 3t)$.

2.2.7. Advanced Model with a Modified Biolek Window Function and a Voltage-Dependent Changeable Integer Exponent

If the memristor voltage signal grows up, then the ionic drift nonlinearity increases as well [4,5,34]. Basically, the representation of the increased nonlinearity of the ionic drift could be illustrated by the decreased integer exponent p in the standard Biolek window function of the memristor [6,34].

The proposed relationship between the positive exponent of the altered Biolek window function p in the suggested memristor model [34] and the absolute value of the memristor voltage signal v could be expressed with a hyperbolic-like dependence [34]—Equation (2.75):

$$p = round \left[\frac{10}{1 + |v|} \right] \tag{2.75}$$

where the special function *"round"* is used for acquiring an integer result [13]. Then, the suggested improved Biolek window function $f_{BM}(x,v)$ applied in the suggested memristor model is Equation (2.76) [34]:

$$f_{BM}(v, x) = 1 - (x - 1)^{[2\,round\,(\frac{10}{1+|v|})]}, \quad v(t) \leq 0$$

$$f_{BM}(v, x) = 1 - x^{[2\,round\,(\frac{10}{1+|v|})]}, \qquad v(t) > 0 \tag{2.76}$$

The suggested memristor model could be completely described with System of Equations (2.77) [34]:

$$\frac{dx}{dt} = k\eta i \left[1 - (x-1)^{2round\left(\frac{10}{|v|+1}\right)}\right], \quad v(t) \leq 0, \quad [i(t) \leq 0]$$

$$\frac{dx}{dt} = k\eta i \left[1 - x^{2round\left(\frac{10}{|v|+1}\right)}\right], \quad\quad\quad v(t) > 0, \quad [i(t) > 0] \quad\quad (2.77)$$

$$v = R\,i = i\left[R_{ON}x + (1-x)R_{OFF}\right]$$

where the third equation in System of Equations (2.77) is the state-dependent Ohm's Law for the memristor element [12,34]. Using Equation (2.77) and the finite difference method [13], a pseudo-code-based algorithm is acquired and applied for the computer investigation of the suggested memristor element [34].

The investigations of the suggested memristor model were made in MATLAB environment [13]. Here, a comparison between several results derived by the use of the standard Biolek memristor model and the suggested altered Biolek model was made. The voltage signal for testing the memristor element is: $v(t) = 0.6\sin\left(2\pi \times 40t - \frac{\pi}{3}\right)$. The time graphs of the memristor voltage v and the corresponding value of the integer exponent p are shown in Figure 52(a,b) for visual observation of the alteration of the window function exponent in accordance to the memristor voltage v. In this case, the integer exponent p changes in a range from 6 to 10 in accordance to the total value of the memristor voltage signal v [34].

The state–flux characteristics of the memristor element, according to the standard Biolek model and the suggested modified model, are presented in Figure 53(a,b) for their visual comparison regarding the form and the corresponding ranges [34]. According to the standard Biolek model, the state–flux relationship is a multi-valued function, while under the same conditions, the state–flux characteristic of the suggested memristor model is approximately a single-valued function which is a benefit of the model proposed by the author. The time graphs of memristor current, according to the standard Biolek model and the suggested model, are shown in Figure 54(a,b) for a comparison of these two models and discussion of the main benefits of the suggested model. For both the memristor models, the electric current has a non-sinusoidal shape due to the memristor nonlinearity [5,34]. The current–voltage characteristics of the memristor element according to the standard Biolek model and the suggested altered Biolek model are presented in Figure 55(a,b). It is obvious that the current–voltage functions are pinched hysteresis loops for both the standard Biolek model and the suggested modified Biolek model, and they almost coincide with one another. For the standard Biolek model, the i–v characteristic is a multi-valued function, while for the suggested altered Biolek model, it is approximately shown by a double-valued curve. The diagrams of the standard Biolek window function f_B (x,v) and the suggested altered Biolek window function $f_{BM}(x,v)$ acquired for different values of the positive integer exponent p and for the altered Biolek model with a voltage-dependent integer exponent are presented in Figure 56(a,b). The range of the state variable x for the modified Biolek model grows up with the rising of the integer exponent p, while at the same time, the range of the window decreases. For integer exponents higher than unity, the standard Biolek window function attains its maximum value. With the increasing of the positive integer exponent in the standard Biolek window, the ionic drift nonlinearity decreases [34].

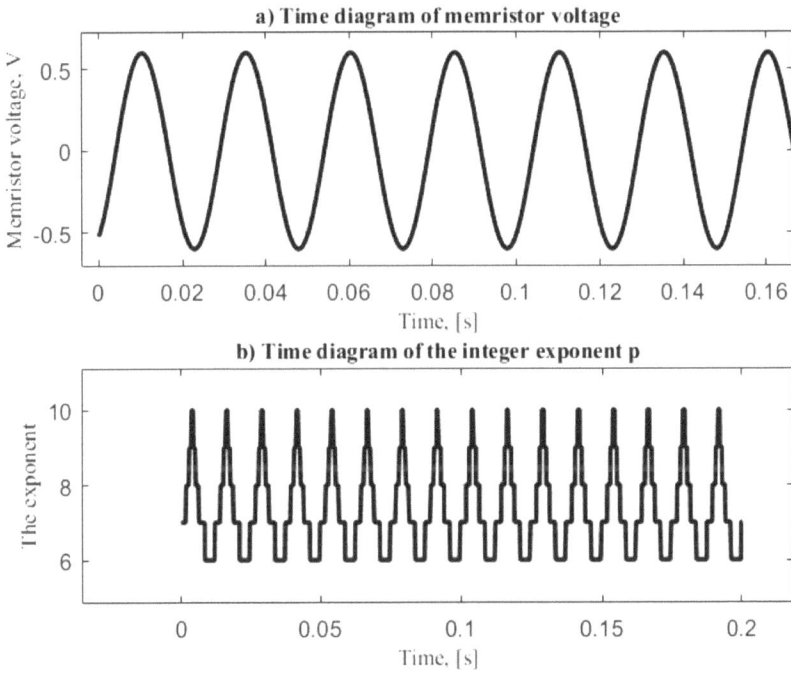

Figure 52. (a) Time diagrams of voltage v and (b) time diagram of the integer exponent p in the altered Biolek window function for the soft-switching regime. The applied voltage is written as: $v(t) = 0.6sin\left(2\pi \times 40t - \frac{\pi}{3}\right)$.

69

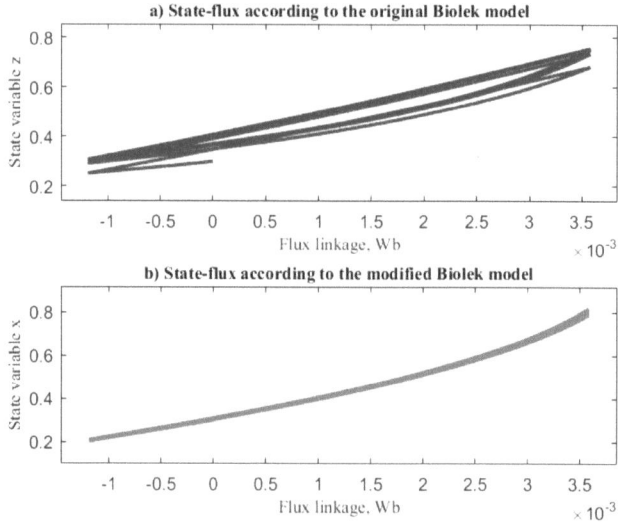

Figure 53. State–flux characteristics of the memristor element according to (**a**) the standard Biolek memristor model; and (**b**) the altered Biolek model for the soft-switching state. The applied voltage is written as: $v(t) = 0.6 \sin\left(2\pi \times 40t - \frac{\pi}{3}\right)$.

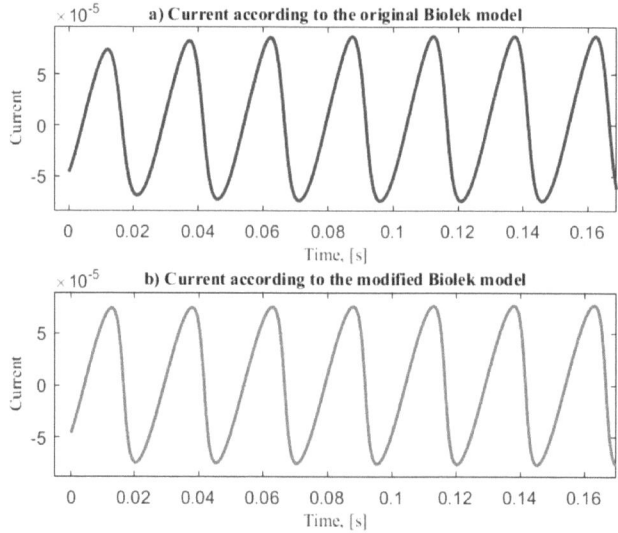

Figure 54. Time graphs of electric current according to (**a**) the standard Biolek model; and (**b**) the altered Biolek model for the soft-switching regime. The applied voltage is written as: $v(t) = 0.6 \sin\left(2\pi \times 40t - \frac{\pi}{3}\right)$.

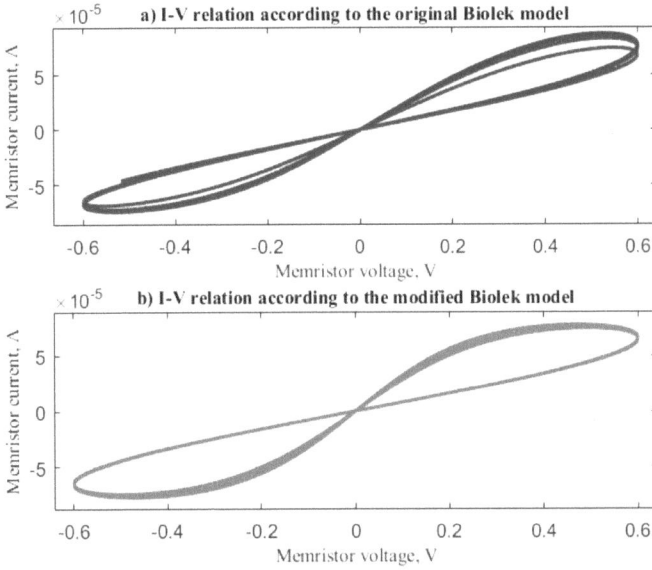

Figure 55. Current–voltage relationships of the memristor according to (**a**) the standard Biolek model; and (**b**) the modified Biolek memristor model for the soft-switching state. The applied voltage is written as: $v(t) = 0.6\sin\left(2\pi \times 40t - \frac{\pi}{3}\right)$.

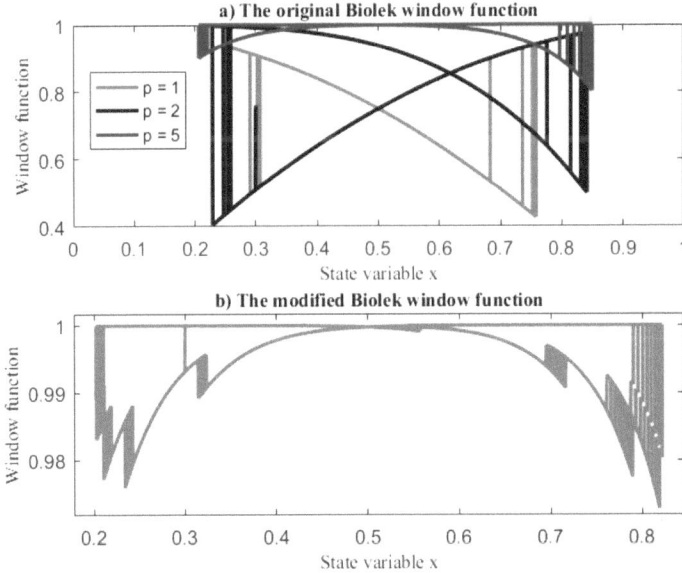

Figure 56. Diagrams of the window functions according to (**a**) the standard Biolek model, and (**b**) the altered Biolek model for the soft-switching state. The applied voltage is written as: $v(t) = 0.6\sin\left(2\pi \times 40t - \frac{\pi}{3}\right)$.

After a comparison of the standard Biolek window functions for different values of the positive integer exponent p with the suggested altered window function, it could be concluded that the altered Biolek window function is created from several fragments taken from the standard Biolek window functions derived for different integer exponents. According to the suggested improved memristor model, the positive integer exponent p changes without human intervention in the simulation process in accordance to the absolute value of the voltage v. The working point of the memristor in the field of the dependence between the improved window function and the state variable x is moving between several different fragments for different values of the positive integer exponent p [34].

The voltage signal applied for the computer analysis in a hard-switching state is: $2\sin\left(2\pi \times 40t - \frac{\pi}{3}\right)$. The diagrams of the voltage and the corresponding integer exponent p of the suggested improved Biolek window function for the hard-switching state are shown in Figure 57 for observation of the change of the window function integer exponent p, in accordance to the memristor voltage signal v [34]. In the case of hard-switching regime, the integer exponent p changes in a broad range (from 3 to 10) than that acquired for the soft-switching state (from 6 to 10), and the corresponding nonlinearity of the dopant drift is higher.

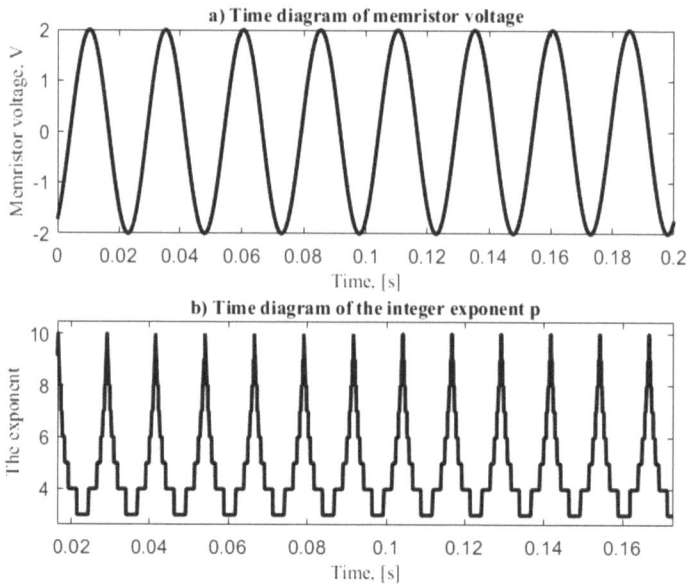

Figure 57. Time graphs of memristor voltage (**a**) and the integer exponent p (**b**) in the altered Biolek window function for the hard-switching regime. The applied voltage is described as: $v(t) = 2\sin\left(2\pi \times 40t - \frac{\pi}{3}\right)$.

The corresponding state–flux characteristics of the memristor for the standard Biolek memristor model [6] and for the suggested modified Biolek model by the author [34] are shown in Figure 58. It is obvious that both the standard Biolek memristor model and the altered Biolek memristor model are capable of restricting the value of the state variable x to the range (0, 1). For the standard Biolek model, the state variable x does not attain the minimum limit of zero, but for the improved memristor model, this limit is almost reached. The diagrams of the electric currents, according to the standard Biolek model and the offered modified Biolek model, are shown in Figure 59(a,b). In both cases, the memristor operates as a rectifier diode [11,34]. In the suggested memristor model, the maximum value of the current i is slightly higher than the corresponding maximum current value for the standard Biolek model [6,34].

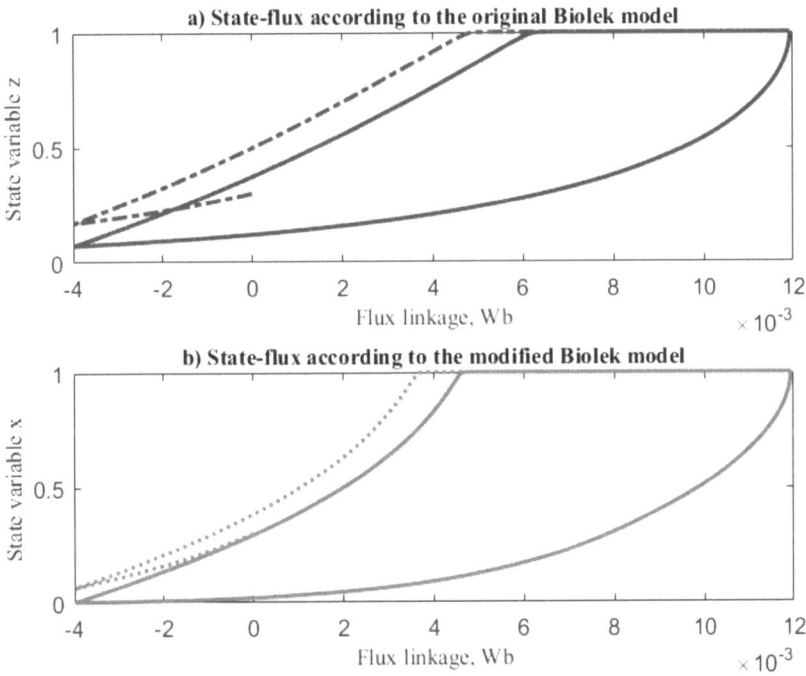

Figure 58. State–flux characteristics of the memristor according to (a) the standard Biolek model and (b) the modified Biolek memristor model for the hard-switching state. The applied voltage is described as: $v(t) = 2\sin(2\pi \times 40t - \frac{\pi}{3})$.

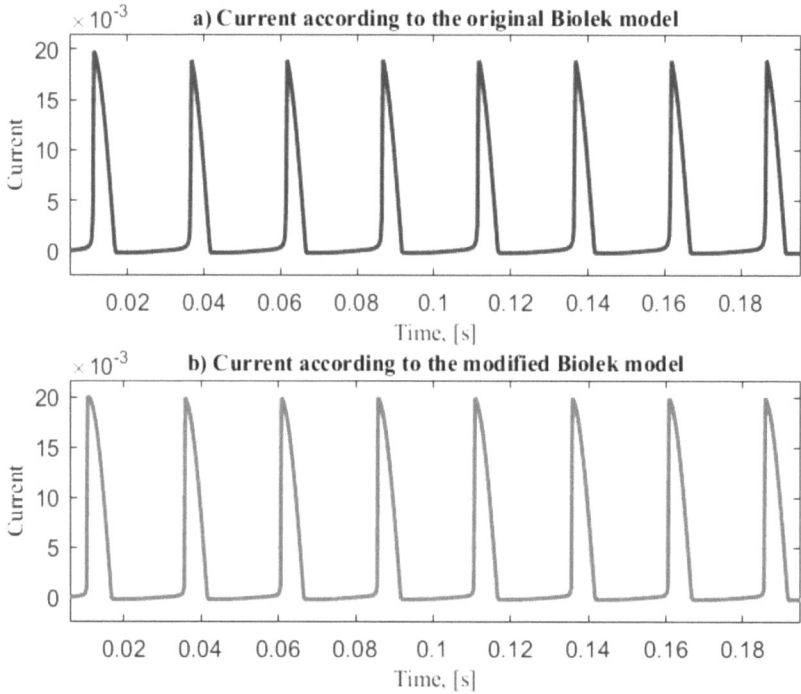

Figure 59. Time graphs of the electric current according to (**a**) the standard Biolek memristor model; and (**b**) the altered Biolek model for the hard-switching state. The applied voltage is described as: $v(t) = 2\sin\left(2\pi \times 40t - \frac{\pi}{3}\right)$.

The current–voltage characteristics of the memristor derived, according to the Biolek memristor model and the altered Biolek model, are shown in Figure 60. In both cases, the current–voltage characteristic is an anti-symmetrical function and confirms the rectifying performance of the memristor when it operates in a hard-switching state [34]. The corresponding current–voltage relationships of the memristor element in both models almost match each other [34]. The diagrams of the standard Biolek window function for different values of the integer exponent p and the altered Biolek window function are given in Figure 61(a,b). For the standard Biolek memristor model, if the integer exponent p is higher than 2, the state variable x and the corresponding window function obtain their maximum ranges, that is, from zero to unity. This phenomenon is also obtained for the suggested altered Biolek memristor model [34].

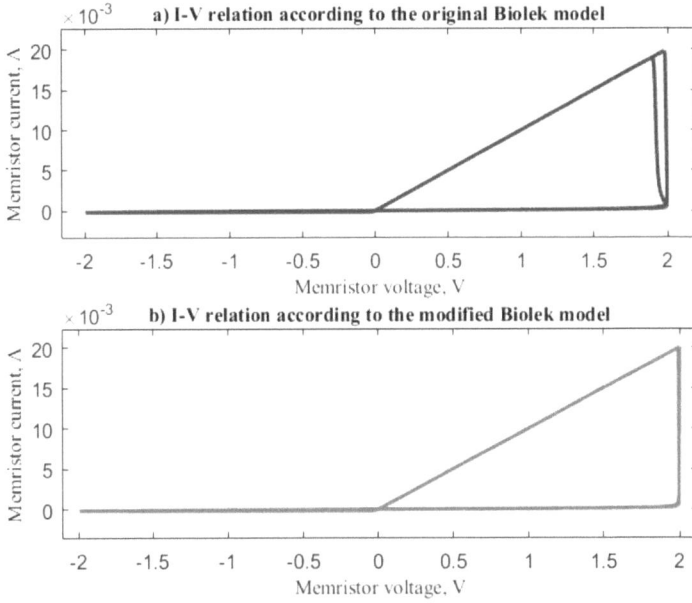

Figure 60. Current–voltage characteristics of the memristor according to (**a**) the standard Biolek model and (**b**) the suggested improved model for the hard-switching state. The applied voltage is described as: $v(t) = 2\sin\left(2\pi \times 40t - \frac{\pi}{3}\right)$.

For the standard Biolek memristor model, the nonlinearity is permanent and the integer exponent p has a value of 1, 2 or 5, while for the improved Biolek model, the nonlinearity depends on the voltage v. Owing to the full range of values of the state variable x (from zero to unity), the hard-switching performance of the memristor element is more remarkably expressed by the altered Biolek memristor model.

For testing the suggested altered memristor model, an Alternating Current (AC) voltage signal with exponentially increasing amplitude is applied. In this case, the transition between the soft-switching regime and hard-switching regime could be visually expressed in the time domain.

The voltage applied for the current computer analysis is as follows: $v(t) = 0.6e^{3t}\sin\left(2\pi \times 40t - \frac{\pi}{3}\right)$. The time graphs of the pseudo-sinusoidal voltage v and the corresponding integer exponent p of the suggested improved Biolek window function are shown in Figure 62 for visual illustration in the time domain. It can be observed the alteration of the positive exponent range is dependent on the used memristor voltage v [34].

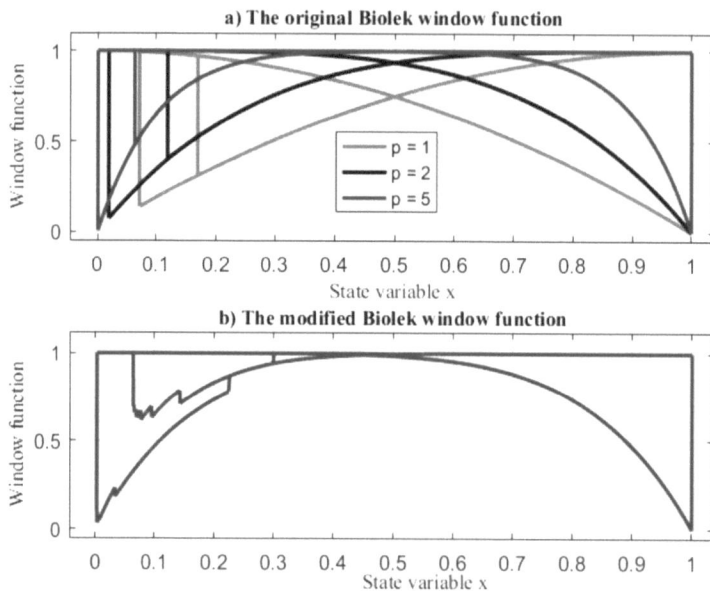

Figure 61. Diagrams of the window functions according to (**a**) the standard Biolek model; and (**b**) the suggested altered Biolek model for hard-switching regime. The applied voltage is described as: $v(t) = 2\sin\left(2\pi \times 40t - \frac{\pi}{3}\right)$.

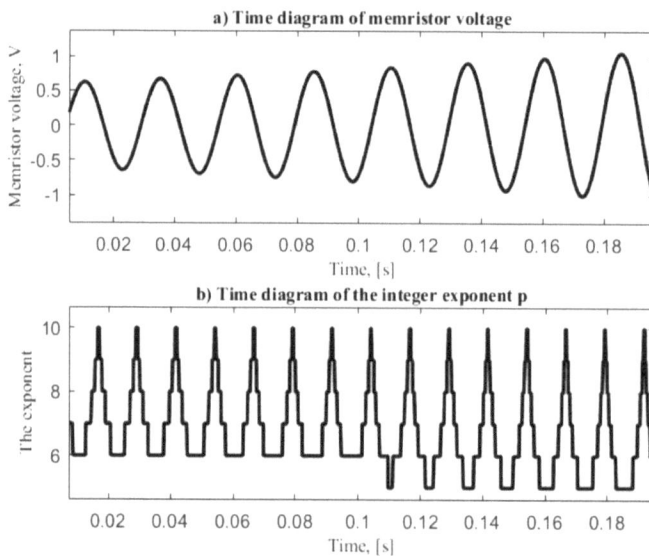

Figure 62. Time graphs of (**a**) the memristor voltage v and (**b**) the positive integer exponent p for a pseudo-sinusoidal voltage with an exponentially increasing amplitude, described as $v(t) = 0.6e^{3t}\sin\left(2\pi \times 40t - \frac{\pi}{3}\right)$.

The state–flux characteristics of the memristor, derived by the standard Biolek memristor model and the altered Biolek model are shown in Figure 63(a,b). In both cases, the state–flux characteristics are acquired as multi-valued hysteresis functions [34]. The time graphs of the memristor currents for the standard Biolek model and the suggested altered Biolek model are shown in Figure 64. For both Biolek models, the current has very low-level values and in the end of the analysis the current amplitude grows up and the memristor starts to operate in a hard-switching state [34].

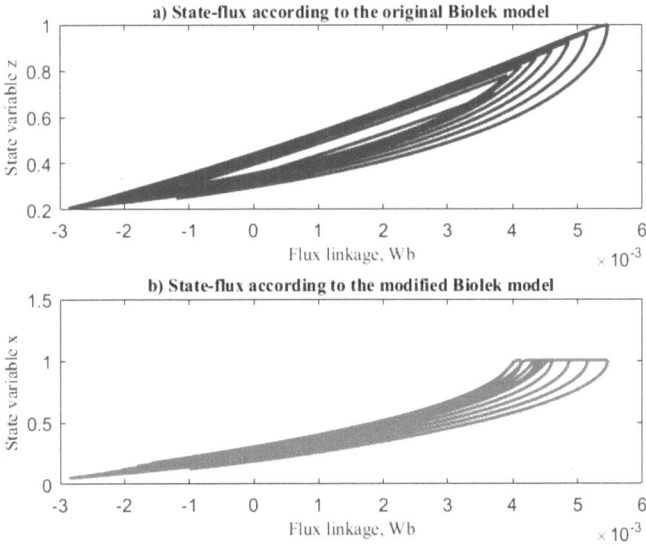

Figure 63. State–flux characteristics of the memristor according to (**a**) the standard Biolek model; and (**b**) the improved Biolek model for a pseudo-sinusoidal voltage with an exponentially increasing amplitude, described as $v(t) = 0.6e^{3t} \sin(2\pi \times 40t - \frac{\pi}{3})$.

For the suggested improved Biolek model, the memristor current is derived with a magnitude several times higher than the current magnitude acquired by the standard Biolek model. The current–voltage characteristics of the memristor element, according to the standard Biolek model and to the offered altered model, for a pseudo-sinusoidal voltage signal with an exponentially increasing magnitude are shown in Figure 65(a,b). Observing the current–voltage characteristics of the memristor element, one can establish the transition between the soft-switching and hard-switching memristor states. In the same environment, the soft-switching behaviour is dominating for the standard Biolek model, while the hard-switching regime is more remarkable for the improved memristor model suggested by the author [34].

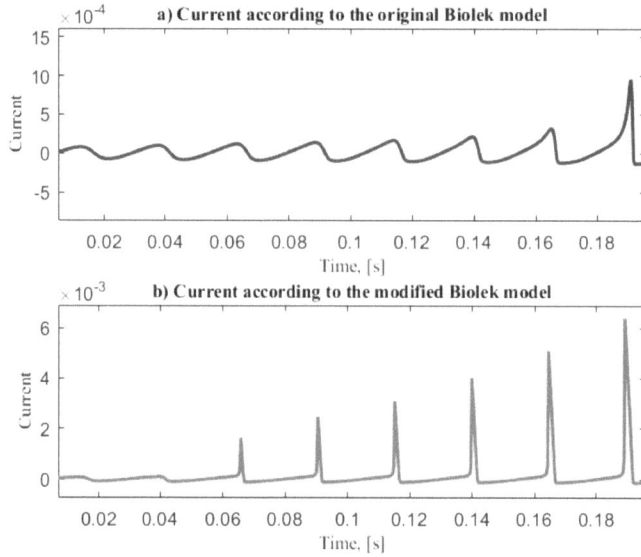

Figure 64. Time graphs of the electric current according to (**a**) the standard Biolek model and (**b**) the altered Biolek model for a pseudo-sinusoidal voltage with an exponentially increasing magnitude, described as $v(t) = 0.6e^{3t} \sin\left(2\pi \times 40t - \frac{\pi}{3}\right)$.

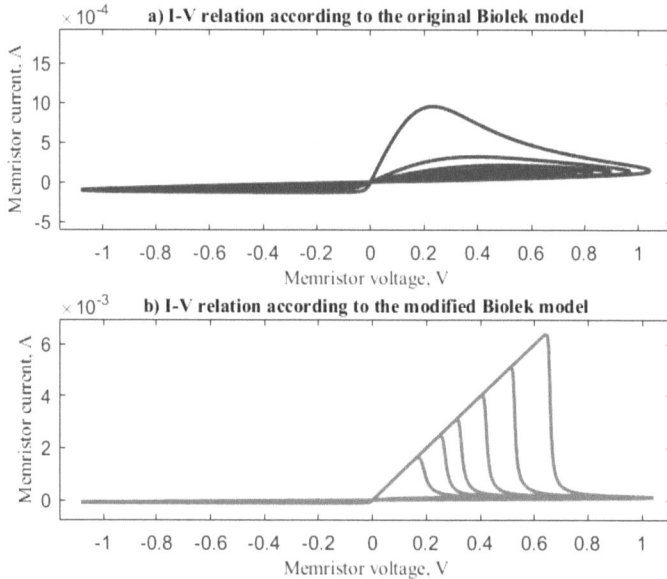

Figure 65. Current–voltage characteristics of the memristor, according to (**a**) the standard Biolek model; and (**b**) the altered Biolek model for a pseudo-sinusoidal voltage with an exponentially increasing amplitude, described as $v(t) = 0.6e^{3t} \sin\left(2\pi \times 40t - \frac{\pi}{3}\right)$.

The window functions for several different positive exponents in the standard Biolek model and the improved memristor model are shown in Figure 66(a,b) for illustration of the trajectories of the working point of the memristor element during the computer analysis.

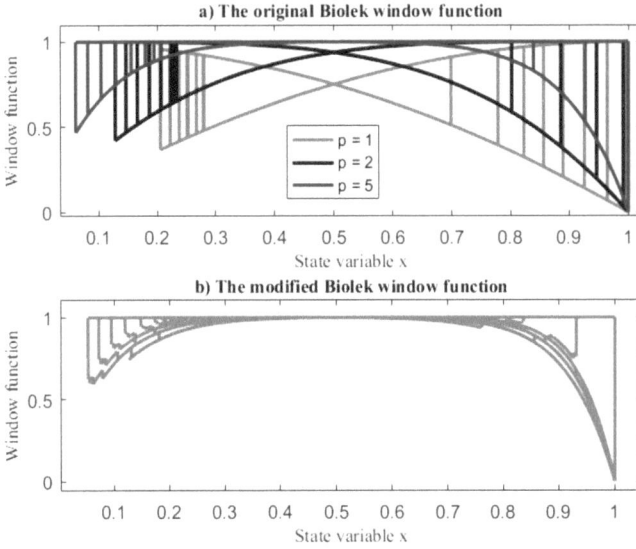

Figure 66. Diagrams of window functions according to (**a**) the standard Biolek model; and (**b**) the modified Biolek model for a pseudo-sinusoidal voltage with an exponentially increasing amplitude, described as $v(t) = 0.6e^{3t} \sin\left(2\pi \times 40t - \frac{\pi}{3}\right)$. If we examine the window functions for the standard Biolek model shown in the first sub-figure of Figure 66(a,b), then it could be established that for the given conditions, the state variable x tends to attain its minimal value of zero when we grow up the integer exponent p.

In this situation, the ionic drift nonlinearity for the altered Biolek memristor model changes in the investigation process, while for the standard Biolek memristor model, the integer exponent has a constant value. The nonlinearity of the dopant drift decreases if a higher value of the integer exponent p is chosen [34].

The suggested improved window function shown in the second sub-figure of Figure 66(a,b) is acquired by the application of several fragments of the standard Biolek window function with different positive integer exponents. This fact is based on the suggested relationship between the integer exponent p and the absolute value of the voltage v [34]. After the comprehensive analytical explanation and the computer investigation of the suggested improved Biolek memristor model, in parallel to the standard Biolek model analysis, several conclusions for the discussed model could be made.

The new altered nonlinear model proposed in this research is based mostly on the standard Biolek model and has many benefits of the standard Biolek model. The suggested model has a nonlinear ionic drift and a mechanism for restriction of the state variable x to the range from zero to unity, established by investigations for hard-switching state.

The new improved model [34] has a benefit, which the standard Biolek model with a constant exponent does not have, that is, the capability of more realistically depicting dopant drift nonlinearity in accordance to the absolute value of the applied voltage v. As it could be observed from the acquired results, the state–flux characteristics of the altered Biolek model for the soft-switching mode are almost single-valued functions. This is an advantage of the proposed model the standard Biolek model lacks, which illustrates, under the same circumstance, multi-valued state–flux characteristics of the memristor [34].

2.2.8. A Model with a Modified Window Function and an Activation Threshold

The proposed model is constructed by the use of Biolek model [6], Joglekar model [5] and GBCM model [14]. The considered model has the main benefits of the described memristor models above. The application of a window function with a voltage-dependent positive integer exponent is the new suggestion in this model. The adapted window function is based on both Biolek and Joglekar window functions [35]. As the GBCM memristor model, the adapted model uses activation threshold [35]. The new suggestion in the altered memristor model is the possibility for changing the integer exponent of the window function and the corresponding nonlinearity of the ionic motion in accordance to the voltage signal v. This mechanism is significant for the rational illustration of the nonlinear ionic flow [35], which is a function of the memristor voltage v. The inspiration for the current investigation is related to the absence of a correlation between the ionic motion nonlinearity and the voltage v in many models, such as GBCM [14], Joglekar [5], and Biolek [6] models. The opportunity for realistic depiction of the nonlinear dopant motion and the fractional improvement of the flux–charge characteristic of the element, in accordance to the Pickett memristor model, is connected to the use of a function between the integer exponent of the window function p [35] and the applied voltage signal v.

The suggested memristor model by the author is based on the Biolek model, Joglekar model and GBCM memristor model [35]. It has their benefits—demonstration of the nonlinear dopant drift in the memristor element according to the state variable x, a constant resistance for signals lower than the sensitivity threshold, and realistic illustration of the boundary effects for the hard-switching regime. The main and interesting new benefit of the suggested model is the realistic depiction of the nonlinear ionic motion, in which nonlinearity is dependent on the applied voltage v and on the corresponding electric field intensity E. For this reason, a simplified hyperbolic-like approximated function between the window integer exponent p and the voltage v is offered [35]. The applied additional hyperbolic function has two variables in the numerator and the denominator. The numerical values of these variables are approximately established after comparison of the results with those acquired by the Pickett model and the following adjustment. The altered window function $f_M(x)$ proposed here is based both on Joglekar [5] and Biolek [6] window functions. It is a simple linear mixture of the described window functions. Its maximum value of the state variable x is unity and the corresponding minimum value is zero. The suggested window function $f_{BJM}(x)$ by the author is given in Equation (2.76) [35]:

$$f(x) = f_{BJM}(x,v) = \frac{f_B(x,v) + f_J(x)}{2} \tag{2.78}$$

After substituting the expressions for Joglekar and Biolek window functions into Equation (2.76), a more suitable expression of the window function is acquired and presented in Equation (2.79):

$$f_M(x) = 1 - \frac{(x-1)^{2p} + (2x-1)^{2p}}{2}, \quad v(t) \le 0$$

$$\tag{2.79}$$

$$f_M(x) = 1 - \frac{x^{2p} + (2x-1)^{2p}}{2}, \quad v(t) > 0$$

If the signal v grows up, the nonlinearity of the ionic drift increases [35]. The depiction of the increased nonlinearity of the ionic motion could be expressed by diminishing the positive integer exponent p in the altered window function [35]. There are many potential expressions for illustrating such a relationship. The author tried to apply a simplified relationship for optimizing the model and reducing the needed computational time. The offered simplified window function between the positive integer exponent p and the memristor voltage v is given in the next Equation (2.80) [35]:

$$p = round\left(\frac{a}{c + |v|}\right) \tag{2.80}$$

where the specialized function "$round$" is applied for deriving an integer outcome [35]; the variables a and c could be approximately established after comparison of the acquired results to those derived by the Pickett memristor model and correction of the modified model; the constant c is used for avoiding the division by zero if the memristor voltage signal v has a value of zero in the simulation process. The adapted window function $f_M(x,v)$ (System of Equations (2.80)) is substituted into Equation (2.79). Applying the state-dependent current–voltage relationship, the offered model could be illustrated with System of Equations (2.79) [35]:

$$
\begin{aligned}
\frac{dx}{dt} &= \eta\, k\, i \left\{ 1 - \frac{1}{2} \left[\begin{array}{l} (x-1)^{2\cdot round\left(\frac{a}{|v|+c}\right)} + \\ +(2x-1)^{2\cdot round\left(\frac{a}{|v|+c}\right)} \end{array} \right] \right\}, \quad v(t) \le 0 \\
\frac{dx}{dt} &= \eta\, k\, i \left\{ 1 - \frac{1}{2} \left[\begin{array}{l} x^{2\cdot round\left(\frac{a}{|v|+c}\right)} + \\ +(2x-1)^{2\cdot round\left(\frac{a}{|v|+c}\right)} \end{array} \right] \right\}, \quad v(t) > 0 \\
v &= R\,i = \left[R_{ON}x + R_{OFF}(1-x) \right] i
\end{aligned}
\tag{2.81}
$$

where the third equation in System of Equations (2.81) presents the state-dependent Ohm's Law [4,12]. The analysis of the memristor element is completed using the numerical solution to System of Equations (2.81) by applying the finite difference method in MATLAB [13]. An algorithm similar to those used in the GBCM model [14] is used here for illustration of the border effects. The offered memristor model is analyzed for the same voltage signal described as $v(t) = 0.5\sin(2\pi \times 4t)$, used for the Pickett model investigation in the PSpice environment [19]. The current–voltage and flux–charge characteristics of the element according to the Pickett model are shown in Figure 67(a,b).

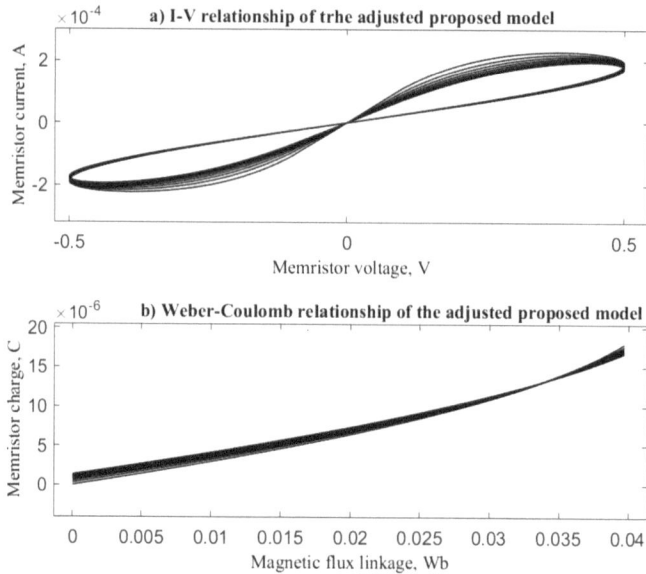

Figure 67. (a) Current–voltage relation of the memristor according to the adjusted altered model; (b) flux-linkage–charge relationship according to the adjusted suggested memristor model. The applied voltage signal is described as: $v(t) = 0.5\sin(2\pi \times 4t)$.

After numerous analyses of the offered memristor model for different values of the variables a and c, it is established that, for $a = 30$ and $c = 2$, the current–voltage and the flux–charge relationships shown in Figure 67(a,b) are approximately identical to those acquired by the Pickett model. Then, the suggested memristor model is tuned according to the Pickett model.

The current–voltage and the flux–charge characteristics of the element are illustrated by multi-valued curves derived in approximately comparable ranges in the fields of the respective coordinates. During the investigations of Pickett memristor model and applying voltage signals higher than 0.7 V, several convergence issues appear [35]. The basic advantage of the suggested memristor model by the author, compared to the Pickett model, is the absence of computational issues.

Investigation of the suggested memristor model for the soft-switching state, in parallel to the GBCM model and the standard Biolek model, is completed for voltage described as $v(t) = 2.55\sin(2\pi \times 4t)$. Its current–voltage characteristics acquired by the computer simulation are shown in Figure 68(a) for representation of its main properties. The current–voltage characteristic of the suggested memristor model is a multi-valued pinched hysteresis loop. After a comparison with experimentally derived current–voltage curves [17,20], a good correspondence between the respective relations is established. The current–voltage function acquired by the GBCM memristor model is a double-valued pinched hysteresis loop due to the linearity of the ionic motion in the present case [35].

The current–voltage characteristic derived by the standard Biolek memristor model for $p = 1$ is a multi-valued pinched hysteresis loop, which has a higher nonlinearity according to the proposed model. The ranges of the electric current for the offered model and for GBCM model are almost similar, while for the Biolek model, this range is narrower. In this situation, the best results can be obtained by the GBCM model owing to the double-valued current–voltage relationship. The corresponding Weber-Coulomb characteristics are illustrated in Figure 68(b). For the suggested memristor model and the GBCM model, their flux–charge relationships are single-valued curves which approximately match each another [35]. The flux–charge characteristic of the memristor, according to the standard Biolek model, is a multi-valued function in a narrower range, according to the coordinate axis. In this case, the suggested memristor model and the GBCM model are almost equivalent to each other according to their behavior [35]. The analysis of the proposed model [35], the GBCM model [14] and the standard Biolek memristor model [6] for the hard-switching state is made for the following voltage signal: $v(t) = 3.5\sin(2\pi \times 0.7t - \frac{\pi}{2})$. In this case, the current–voltage characteristics presented in Figure 69(a) and derived by the memristor models under test are almost identical and practically match each other. They are similar to the i–v relationship of a semiconductor diode. The flux–charge characteristics of the memristor element are illustrated in Figure 69(b). The flux–charge relationships derived by the suggested model and the GBCM model are almost similar. These curves describing the flux–charge relationships have a hysteresis shape owing to the border effects. The charge of the memristor reaches its limiting value [35]. According to the standard Biolek model, the memristor charge does not reach its minimum value. In this situation, the behaviors of the suggested model and GBCM model are almost identical. The Biolek model acquires high-quality results. The modified model is tested for a signal with a magnitude lower than the activation threshold $v_{thr} = 0.1$ V, and is established that the state variable x in this case is a constant, equal to its initial value x_0. The memristor element behaves as a linear resistor. During the simulations of the suggested model [35], no convergence issues have been observed.

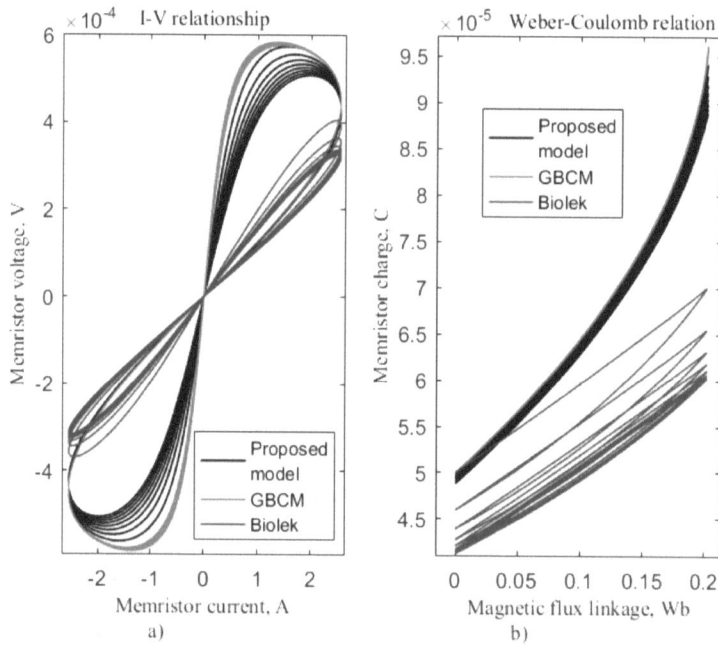

Figure 68. (a) Current–voltage relationships of the proposed model, the GBCM model and the standard Biolek model with $p = 1$ for the soft-switching state; (b) Weber–Coulomb characteristics of the proposed model, the GBCM model and the standard Biolek memristor model with $p = 1$ for the soft-switching regime. The voltage signal for analysis is described as: $v(t) = 2.55 \sin(2\pi \times 4t)$.

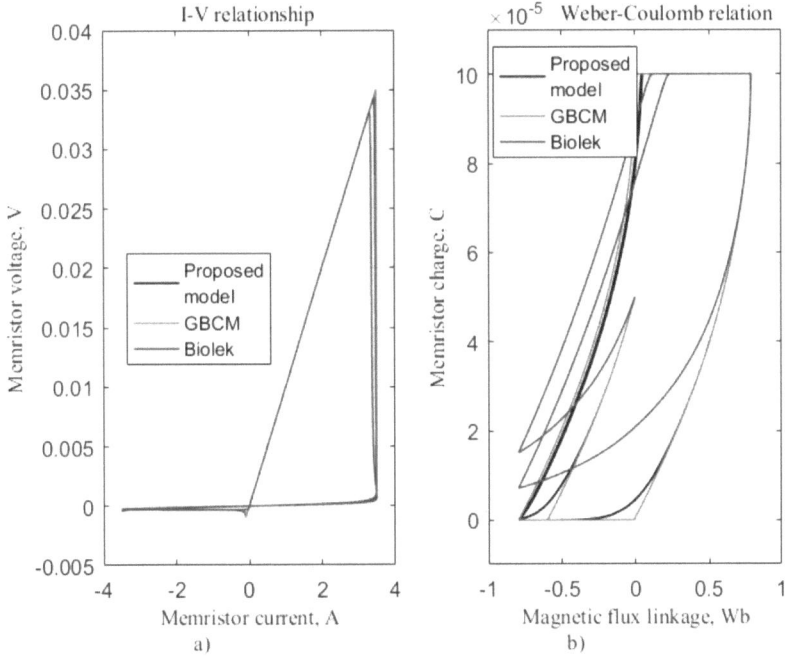

Figure 69. (a) Current–voltage relationships, derived by the proposed model, GBCM model and Biolek memristor model for the hard-switching mode; (b) The corresponding Weber–Coulomb relationships derived by the proposed model, GBCM model and Biolek model. The voltage signal for analysis is described as: $v(t) = 3.5 \sin\left(2\pi \times 0.7t - \frac{\pi}{2}\right)$.

The time diagrams of the window function integer exponent p, the state variable x of the memristor element and the window function $f(x)$ in dependence on the state variable x are presented in Figure 70(a–c) for clarification of the main features of the proposed model for the hard-switching mode [35]. The integer exponent p in the modified window function changes in accordance to the applied voltage v. This integer exponent changes in the range between 6 and 15. The state variable x reaches its limiting values. The modified window function contains several segments. The memristor operating point moves on the window function curve in accordance to the applied flux linkage Ψ [35].

The needed computational times for the suggested model, GBCM model and standard Biolek model are 0.39 s, 0.28 s and 0.15 s, respectively. After a comparison of GBCM memristor model and Biolek model, it could be concluded that the performance of the suggested model is closer to that of the GBCM model [14,35]. The suggested memristor model is tested in a real electronic scheme [35]. For this goal, the integrator given in Figure 71 [35] is principally suitable because it uses two processes—writing information in the memristor by changing its conductance and using an input voltage signal higher than the memristor sensitivity threshold, and reading the stored information by a very small DC current lower than the activation threshold of the element. Several basic models, such as Strukov and Williams's, Joglekar, Pickett and Biolek models are not suitable to be used in such integrators, because they do not have an activation threshold. The integrator device presented in Figure 71 has been simulated in MATLAB environment [13] using the suggested model by the author and the GBCM model [14]. The operation of the proposed integrator device is based on transformation of the time integral of the input voltage signal in electric charge, stored in a memristor [35]. For avoiding the violation of the integration and the related boundary effects, the memristor is operating in a soft-switching regime for low input voltages [35].

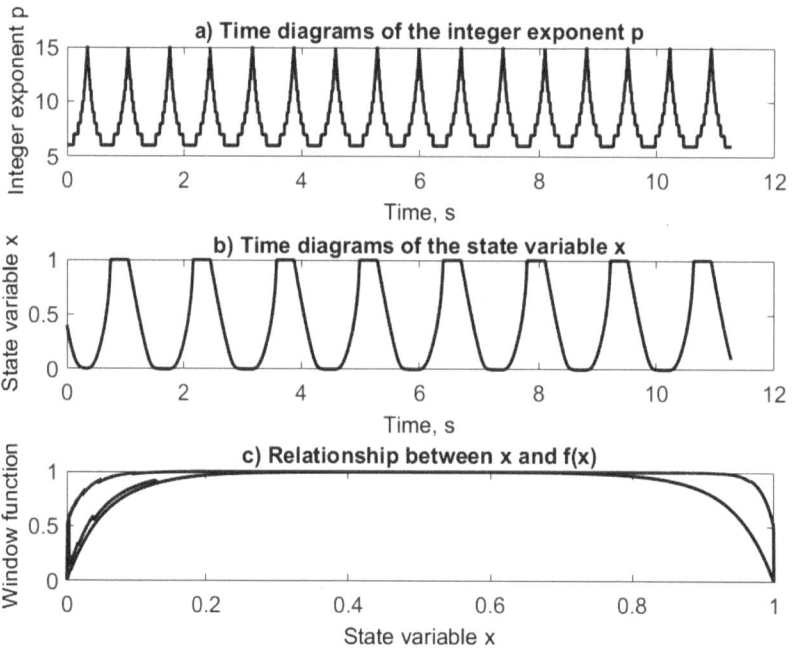

Figure 70. (a) Time diagram of the window function integer exponent p in dependence on the applied voltage v; (b) time diagram of the memristor state variable x; representation of the modified window function, derived for the hard-switching mode; the window function contains several segments, the operating point of the memristor moves on the window function in accordance to the applied flux linkage Ψ.

The information for the resistance of the memristor is read by a DC signal acquired by the current source with a current of 30 μA. This value is lower than the activation threshold of the memristor element [35]. The voltage across the current source and the capacitor is proportional to the resistance of the memristor element [12,35]. The switches sw_1, sw_3 and sw_2, sw_4 are controlled in an anti-phase regime. They are applied for sampling the input signal and for separating the signals in the time domain, to avoid parasitic communication between the electric sources [12,35]. The capacitor C with a capacitance of 560 pF is applied for filtering and smoothing the output voltage signal. The memristor models applied for the simulations have equal values of their basic parameters. The time diagrams of the control signals for the electronic switches are presented in Figure 72(a,b) for visual expression of their anti-phase operating mode. According to the applied control signals, if sw_1 and sw_3 are open, in the same time, s_2 and s_4 are closed, and vice versa [35]. The time diagram of the memristor state variable x and the memristance are presented in Figure 73(a,b) for visual observation of the change of memristor state in time domain. It is observable that both the state variable x and the resistance of the memristor element M change piece-wise linearly and the memristance is proportional to the time integral of the input voltage v. The state–flux and the current–voltage characteristics are shown in Figure 74(a,b). The time diagrams of the input and output voltages of the considered integrator, derived from the present analysis, are given in Figure 75(a,b). The input voltage is a sequence of rectangular pulses with different polarities [35]. The output voltage is proportional to the time integral of the input voltage. When the input voltage is a positive signal with a constant value, the corresponding output signal is illustrated by a monotonically increasing straight line. When the input voltage changes its polarity, the output voltage is shown by a monotonically decreasing straight line. The graphics of the output voltage according to the suggested model and GBCM model match each another. The computational time for GBCM model is 0.31 s and for the offered model it has higher value—0.33 s. For analysis of electronic devices such as the integrator illustrated above, only the GBCM and the proposed model are appropriate due to the use of sensitivity threshold. Both the GBCM and the proposed memristor models derive realistic results [35].

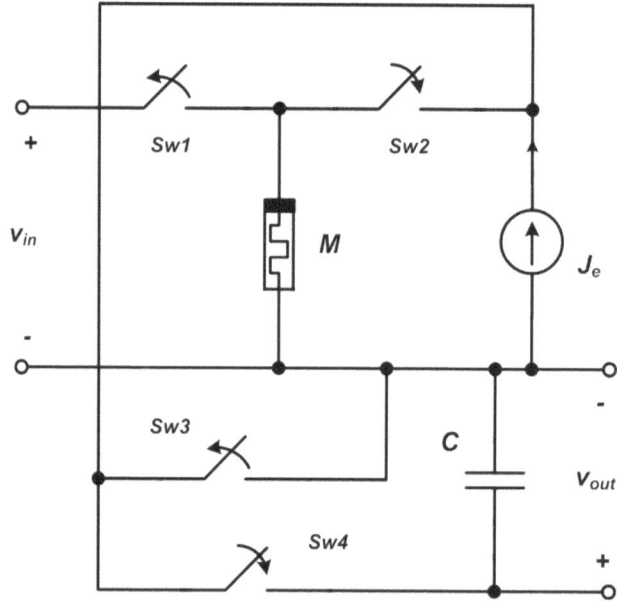

Figure 71. Memristor integrator for testing the suggested memristor model.

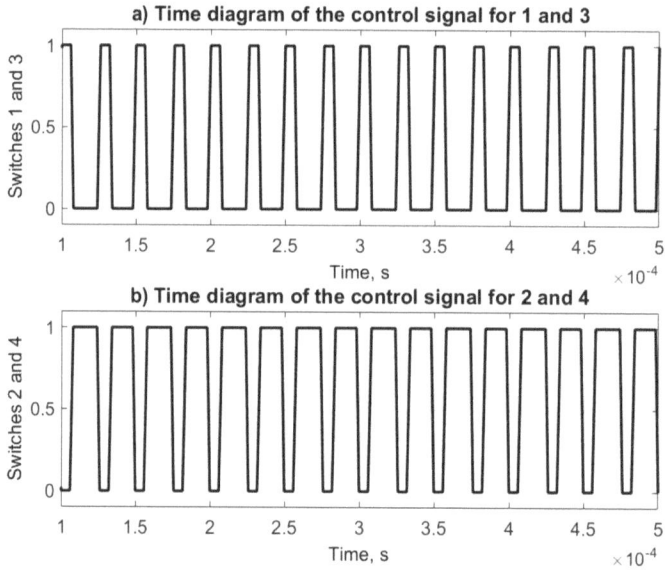

Figure 72. Control signals for the electronic switches in the memristor integrator: (a) Control signal for the switches 1 and 3; (b) Controlling signal for the switches 2 and 4.

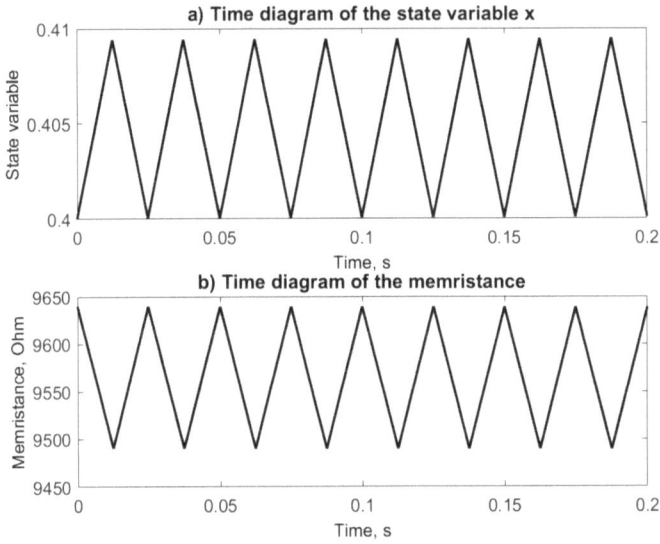

Figure 73. Time diagrams of (**a**) the state variable x and (**b**) the memristance M of the memristor element in an operation mode.

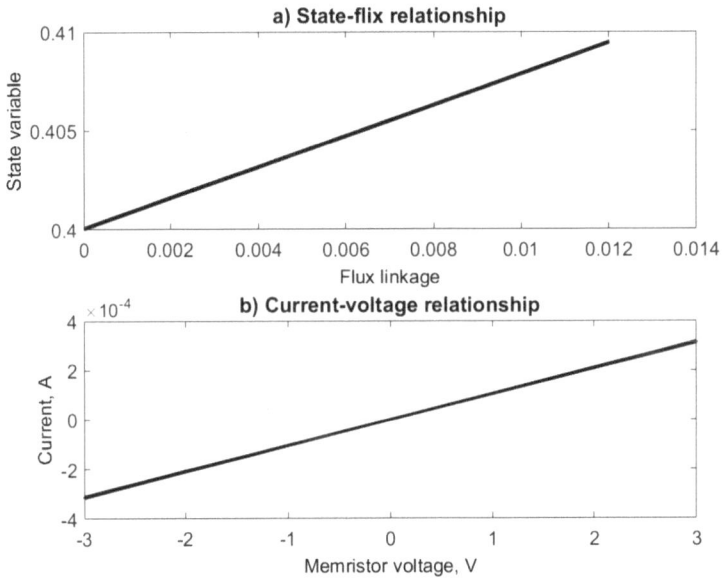

Figure 74. (**a**) State–flux and (**b**) current–voltage characteristics of the memristor element, derived in the normal working mode.

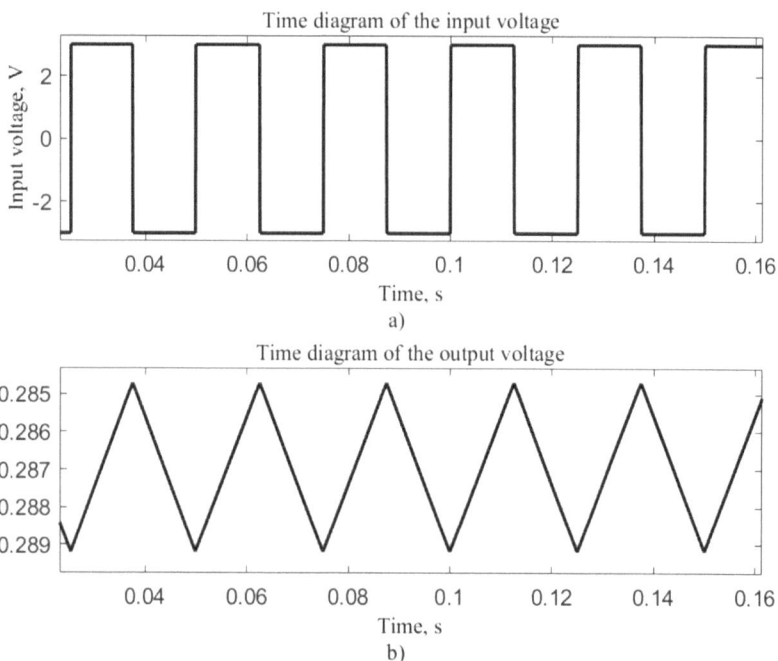

Figure 75. Graphs of the input voltage (**a**) and output voltage (**b**) acquired for the operating impulse regime of the memristor integrator.

The memristor model suggested here is based on several main models, i.e., the Joglekar memristor model, Biolek model and GBCM model. It has their main properties and benefits [35]. It has an activation threshold, a switch-based algorithm for illustration of the boundary effects and the capability for representation of the nonlinear ionic drift. Another advantage of the suggested memristor model, according to the Pickett memristor model, is the established absence of convergence issues. The proposed model has a new improvement compared to the previous described models—the capability for realistic depiction of the ionic drift nonlinearity in accordance to the voltage [35]. For this purpose, the modified window included in the suggested model has a voltage-dependent integer exponent. The suggested memristor model has been effectively tested in a switch-based integrator device in parallel with the GBCM model. The results derived by these two models are almost identical [35].

2.2.9. Summary of the Described Window Functions

In Table 3, the basic described existing and modified window functions by the author are summarized. They could be applied for different types of memristors with equations, similar to th se illustrating the titanium dioxide memristor elements.

Table 3. Summary of several existing and the modified window functions by the author.

Name	Reference of Description	Reference of Application	Mathematical Expression	Cases for Use of the Window Function, when the Advantages are Especially Important	Advantages	Disadvantages	Applicability
Joglekar	[5,11]	[32,35]	$f_J(x) = 1 - (2x - 1)^{2p}$	Soft-switching mode, when the representation of the nonlinear ionic drift is important	Simple expression, low complexity, middle accuracy	Could not represent the boundary effects, missing activation thresholds	Oscillators, memristor-based filters
Biolek	[6,11]	[33,35]	$f_B(x, i) = 1 - \|x - stp(-i)\|^{2p}$ $stp(i) = \begin{cases} 1, & if\ i \geq 0 \ (v \geq 0) \\ 0, & if\ i < 0 \ (v < 0) \end{cases}$	Hard-switching and soft-switching mode, high-level signals, applicable when the representation of boundary effects is very important	Expressing successfully the boundary effects, low complexity, good accuracy	Middle nonlinearity, missing activation thresholds	Memristor-based Memories, neural networks, Digital devices
BCM	[11]	[16]	$f_{BCM}(x) = 0,\ if\ (x = 0\ \&\ v < 0)\ or$ $(x = 1\ \&\ v > 0)$ $f_{BCM}(x) = 1,\ if\ (x \in (0,1))\ or$ $(x = 0\ \&\ v > 0)\ or\ (x = 1\ \&\ v < 0)$	Hard-switching and soft-switching mode, applicable when the representation of the boundary effects is very important	Expressing successfully the boundary effects, low complexity, good accuracy	Linear ionic drift	Memories, neural networks, digital circuits
Modified Joglekar with a voltage-dependent exponent	[32]	[36,37]	$f_J(x) = 1 - (2x - 1)^{2p},\ \|v\| \geq v_{thr}$ $p = round\left(\frac{a}{c + \|v\|}\right)$ $f_J(x_0) = f_J(x_0) = const,\ \|v\| < v_{thr}$	Hard-switching and soft-switching mode, representation the boundary effects, the activation thresholds and the nonlinear ionic drift, applicable when the representation of the memristor nonlinearity with respect to the state variable is very important (filters, generators, memory elements, neural networks)	Expressing successfully the boundary effects (included algorithm for the boundary effects), good accuracy	Middle complexity	Memories, neurons, digital devices, analog memristor circuits, with realistic description of the ionic drift
Modified Biolek with a voltage-dependent exponent	[34]	[34,35]	$f_{BM}(v, x) = 1 - (x - 1)^{2\,round\left(\frac{10}{c+\|v\|}\right)},\quad v(t) \leq 0$ $f_{BM}(v, x) = 1 - x^{2\,round\left(\frac{10}{c+\|v\|}\right)},\quad v(t) > 0$	Hard-switching and soft-switching mode, without sensitivity threshold, applicable when the representation the boundary effects and the nonlinear ionic drift with respect to the state variable are important (filters, generators, memories)	Precise representation the nonlinear ionic drift according to the state variable x	Middle complexity, absence of sensitivity thresholds	Memristor-based integrator, digital and analogue memristor devices, accurate ionic drift description
Modified Biolek with additional sine-wave component	[33]	[37,38]	$f_{BM}(x) = \left[1 - \frac{(x-1)^{2p} + m(\sin^2(\pi x))}{m + 1} \right],\quad v(t) \leq -v_{thr}$ $f_{BM}(x) = \left[1 - \frac{x^{2p} + m(\sin^2(\pi x))}{m + 1} \right],\quad v(t) > v_{thr}$ $f_{BM}(x_0) = f_{BM}(x_0) = const,\quad -v_{thr} < v(t) < v_{thr}$	Hard-switching and soft-switching state, using an activation threshold for low-amplitude signals, applicable when the representation of increased nonlinearity of the ionic drift according to the state variable is important (memories, digital devices, neural networks)	High nonlinearity, accurate representation the boundary effects and the ionic dopant drift according to the state variable x	Middle complexity	Memory crossbars, neurons, analog and digital circuits, representation of highly nonlinear ionic dopant drift
Modified Joglekar-Biolek with a voltage-dependent exponent	[35]	[35,39]	$f_M(x) = 1 - \frac{(x-1)^{2p} + (2x-1)^{2p}}{2},\quad v(t) \leq -v_{thr}$ $f_M(x) = 1 - \frac{x^{2p} + (2x-1)^{2p}}{2},\quad v(t) > v_{thr}$ $f_M(x_0) = f_M(x_0) = const,\quad -v_{thr} < v(t) < v_{thr}$ $p = round\left(\frac{a}{c + \|v\|}\right)$	Hard-switching and soft-switching regime, applying a sensitivity threshold for low-level signals, applicable especially where high accuracy of the representation of the nonlinear ionic drift according to the state variable is very important (memristor-based integrator using sensitivity thresholds, filters, memories, neural networks)	Realistic representation the nonlinear ionic drift according to the state variable x, the activation thresholds and the boundary effects	Middle complexity	Neural networks and memristor-based memories, memristor switch-based integrators, realistic representation of the ionic drift according to the state variable

The main results, the advantages and the behavior of the proposed four memristor models by the author are summarized in Table 3. Their applications and performances in memristor-based electronic circuits are also discussed. The suggested memristor models by the author with modified window functions are based mainly on the three previously described classical memristor models—Joglekar, Biolek and BCM models, and have their basic properties and advantages. The main new advantages of the suggested memristor modified models by the author are related to the realistic representation of the ionic dopant drift, dependent on the applied voltage and the increased nonlinearity, especially in the memristor model [33]. The increased nonlinearity of the suggested memristor models is presented according to the memristor state variable. The improved representation of the nonlinear dopant drift is also shown according to the state variable of the memristor element. Another advantage of the proposed models is the improved similarity between the derived current–voltage relationship and the corresponding experimental current–voltage characteristics. Almost all of the suggested models, except the one in Reference [34], use activation threshold, giving them the possibility for application in devices which have to differentiate high-level signals from low-level signals, for instance, the memristor-based memory crossbars [38,40] introduced in Chapter 4 and the switch-based integrator [35].

CHAPTER III

Investigation of Memristor Circuits and Devices

3.1. Investigation of a Series Circuit with Two Memristor Elements for the Sinusoidal and Pulse Regime

3.1.1. General Information

The series circuit proposed here for analysis contains two equivalent memristor elements connected in a series connection [41]. The circuit is supplied by a sinusoidal voltage source. The purpose of the present analysis is to compare the behavior of the memristor circuit and the single memristors, and obtain their basic parameters, properties and characteristics. The electrical circuit of the considered scheme is presented in Figure 76 for further comments [41].

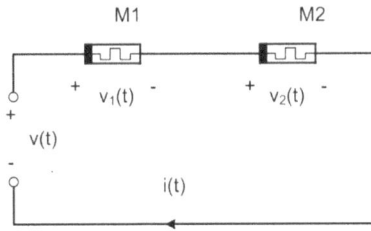

Figure 76. A series circuit with two memristor elements.

3.1.2. Mathematical Description of the Memristor Circuit

For the current analyses, the linear dopant drift model is applied [3,4,41]. It could be briefly described by System of Equations (3.1):

$$\left| \begin{array}{l} v = i\left[R_{ON}x + R_{OFF}(1 - x)\right] \\[2mm] \frac{dx}{dt} = k\,i \end{array} \right.$$

(3.1)

where the first equation is the state-dependent Ohm's Law and the second one is the state differential equation [3,4]. The voltage drops across the two memristor elements and their state differential equations are as follows—System of Equations (3.2) [41]:

$$\left| \begin{array}{ll} v_{M1} = \left[(R_{ON} - R_{OFF})x_1 + R_{OFF}\right]i, & \frac{dx_1}{dt} = k\,i \\[2mm] v_{M2} = \left[(R_{ON} - R_{OFF})x_2 + R_{OFF}\right]i, & \frac{dx_2}{dt} = k\,i \end{array} \right.$$

(3.2)

After expression of the current and using Kirchhoff's Voltage Law [12], the following current–voltage relationship of the whole memristor circuit is derived–Equation (3.3):

$$v = \left[(R_{ON} - R_{OFF})(x_1 + x_2) + 2R_{OFF}\right]i$$

(3.3)

Using System of Equations (3.2) and Equation (3.3), the state differential Equation (3.4) of the whole memristor circuit is acquired:

$$||(R_{ON} - R_{OFF})(x_1 + x_2) + 2R_{OFF}]d(x_1 + x_2) = 2kv(t)dt \tag{3.4}$$

The equivalent resistance of the considered memristor circuit is expressed by Equation (3.5):

$$R_{eq} = \frac{v}{i} = (R_{ON} - R_{OFF})(x_1 + x_2) + 2R_{OFF} \tag{3.5}$$

3.1.3. Results Derived by the Analysis of the Memristor Circuit

The time diagrams of the applied memristor voltage, the current and the resistance of the memristor for the hard-switching mode are presented in Figure 77(a) for establishing several basic memristor properties. The voltage signal used for the hard-switching mode is: $v(t) = 7\sin(2\pi \times 70t)$ [41].

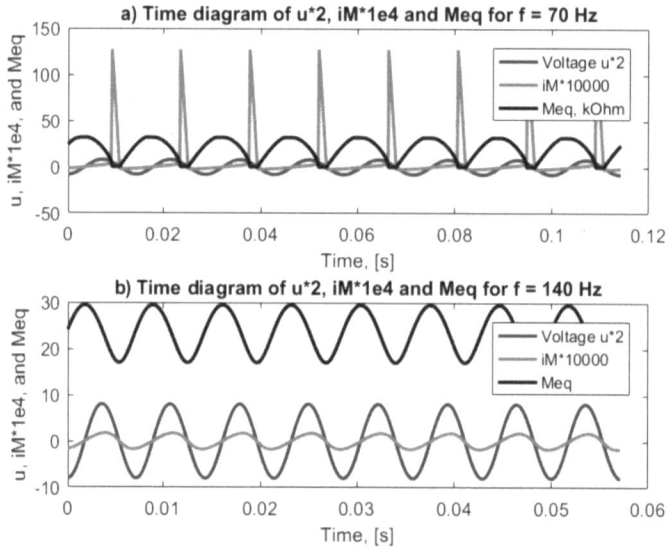

Figure 77. (a) Time diagrams of the current, source voltage, and the equivalent resistance of the memristor circuit, for hard-switching; (b) Time diagrams of the current, source voltage, and the equivalent resistance for soft-switching modes, respectively, which demonstrates that the memristor circuit has a single-memristor behavior.

Similar time diagrams of the applied memristor voltage, the current and the resistance of the memristor are presented in Figure 77(b) for the soft-switching mode. The voltage signal used for the soft-switching mode is: $v(t) = 7\sin(2\pi \times 140t)$. For the hard-switching mode, the memristance changes in the whole range, and the current only has positive values, due to the rectifying effect of the memristor operating in a hard-switching mode. For the soft-switching mode, the memristance of the whole circuit changes in a narrower range than that for the hard-switching mode, and does not reach its limiting values. In this case, the series memristor circuit does not have a rectifying effect and the current has positive and negative values [3,41].

On a basis of the results for hard-switching and soft-switching modes, it could be concluded that the series memristor circuit behaves like a single memristor element. The time diagrams of the state variables of the memristors x_1 and x_2 are presented in Figure 78(a,b) for observation of their behavior in the time domain. The initial values of the state variables x_1 and x_2 are different.

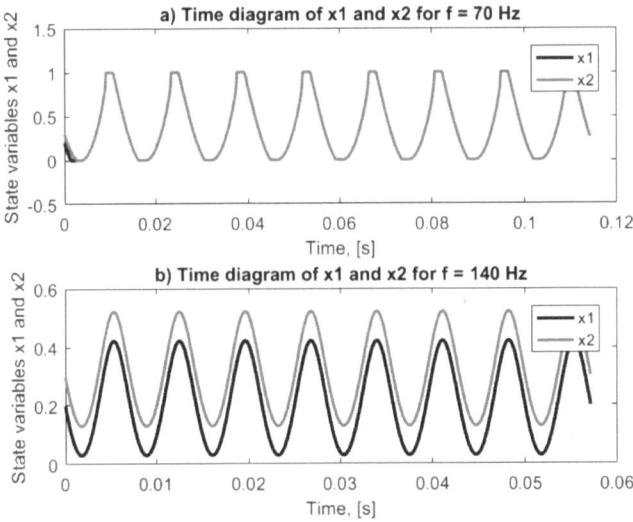

Figure 78. Time diagrams of the state variables x_1 and x_2 of the memristors for (a) hard-switching and (b) soft-switching modes. For the hard-switching mode, the two graphs the state variables x_1 and x_2 match each other; for the soft-switching mode, the two graphs do not exactly coincide, due to their initial values.

For the hard-switching mode, these time diagrams of the state variables x_1 and x_2 almost match each other, while for the soft-switching mode, they do not coincide, but are very close to each other [41]. The state–flux relationships of the two memristor elements for hard-switching and soft-switching modes are presented in Figure 79 for description of their behavior and the memristor properties. For the hard-switching mode, the state–flux relationships of the two memristor elements due to the boundary effects are illustrated by hysteretic curves. For the soft-switching mode, the state–flux relationships of the two memristor elements are expressed by monotonically increasing single-valued curves. The respective current–voltage characteristics of the two memristor elements are presented in Figure 80 for expression of their shapes and ranges for hard-switching and soft-switching modes. For the hard-switching mode, the current–voltage relationships of the memristor elements almost match each other and the memristor elements have rectifying behavior. For the soft-switching mode, the respective current–voltage relationships of the two memristor elements are pinched hysteresis loops, which are very close to each other [41].

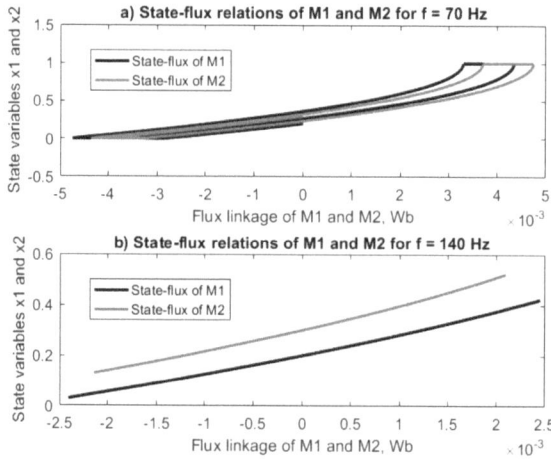

Figure 79. State–flux relationships of the memristors for (**a**) hard-switching and for (**b**) soft-switching modes, representing the behavior of a single memristor for these specific modes.

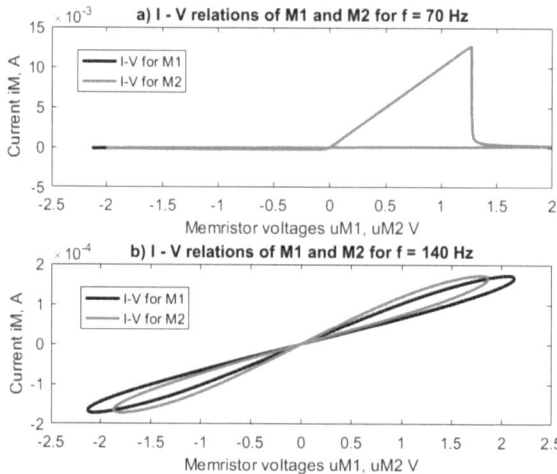

Figure 80. (**a**) Current–voltage relationships of the two memristor elements in the series memristor circuit for hard-switching and (**b**) for soft-switching modes.

The equivalent current–voltage relationships of the whole memristor circuit for hard-switching and soft-switching modes are presented in Figure 81 for expression of the behavior of the whole memristor circuit. Based on the respective current–voltage relationships of the two memristor elements, it could be established that the series circuit with two memristors behaves like a single memristor [41]. The equivalent resistance of the circuit is higher than the resistance of a single memristor. The respective harmonic components in the frequency spectrum of the memristor current for hard-switching and soft-switching modes are presented in Figure 82(a,b). It is observable that the frequency spectrum for the hard-switching mode is many times richer than the frequency spectrum for the soft-switching mode [41].

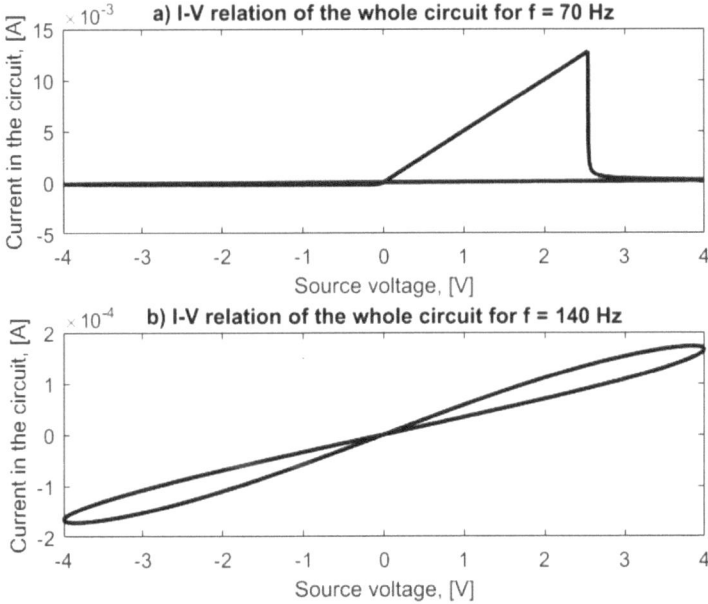

Figure 81. (a) Equivalent current–voltage relationships of the whole circuit for hard-switching and (b) for soft-switching modes.

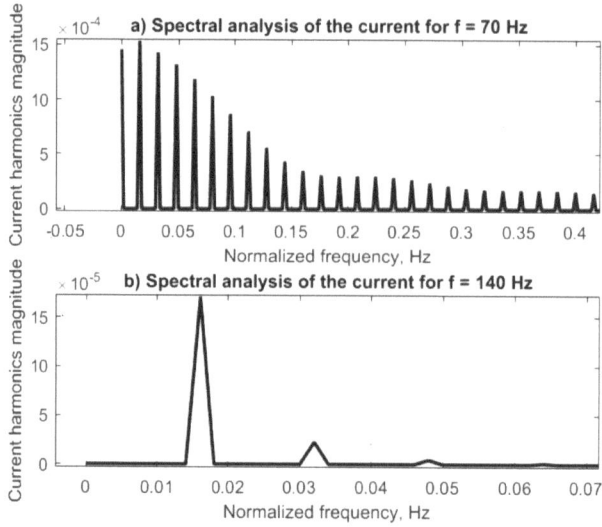

Figure 82. Spectral analyses of the current for (a) hard-switching and (b) for soft-switching modes.

3.1.4. Discussion of the Derived Results

The behavior of the series memristor circuit is similar to that of the single memristor element. The resistance of the entire circuit is higher than that of a single memristor. The states of the included memristor elements depend on the initial values of the state variables and on the applied voltage.

3.2. Investigation of a PSpice Model of a Titanium dioxide Memristor Element and a Memristor-Based Wien Oscillator

3.2.1. General Information for the Analyzed Circuit

A significant property of the memristor is that its memristance could be simply regulated by applying electrical impulses. Besides this, in some segments of its current–voltage relationship, the memristor element has a negative differential resistance [3,4,12,42]. The discussed memristor properties are prerequisite for its potential application in generator schemes [12,42]. The main goals of this analysis are synthesis of a suitable PSpice [19] memristor model and explanation of its application for investigation of a memristor-based Wien oscillator [19,42]. For the current investigations, the current–voltage correlation for linear ionic dopant drift, expressed by Equation (3.6) is applied [12,42]:

$$i(t) = \frac{v(t)}{R_{OFF}\sqrt{\left(1 - \frac{q(t_0)}{Q_D}\right)^2 - \frac{2\eta}{Q_D R_{OFF}}\int v(t)dt}} \tag{3.6}$$

100

The circuit of the investigated memristor-based Wien oscillator is shown in Figure 83 [12,42]. It is derived by alteration of the classical scheme of the generator with Wien bridge by substituting several resistors with memristors [42]. The operational amplifier applied in the circuit operates as a master nonlinear unit [12,19,42]. The resistances of the memristor elements applied in the circuit could be changed by external sources as presented in Figure 83 [12,42]. The deviation of the resistances of the memristor elements M_1 and M_2 is used for controlling the frequency of the output signal of the generator device. By altering the equivalent resistance of the memristor elements M_3 and M_4, the amplitude of the output signal could be altered in a predetermined time interval [12,42].

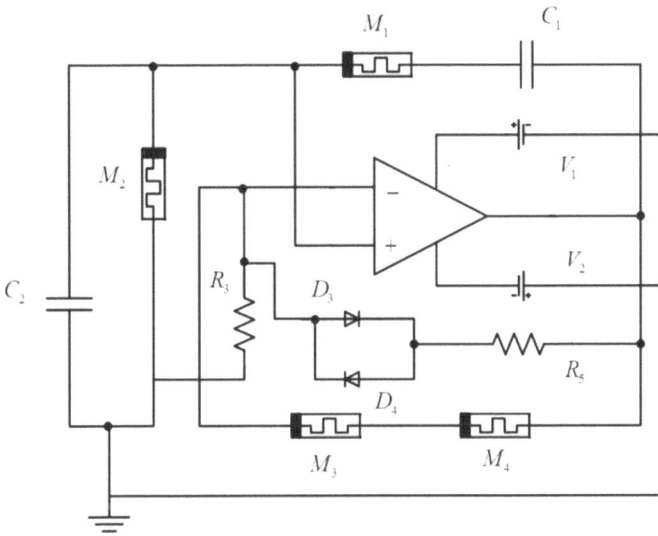

Figure 83. A memristor-based Wien generator scheme, representing its basic structure and principle of operation in oscillating mode.

The function of the diodes D_3 and D_4 is to stabilize the amplitude of the output voltage when the ambient temperature changes. The Wien bridge consists of two capacitors (C_1 and C_2) with equal capacitances of 4.7 nF and two memristor elements M_1 and M_2 with equal values of the initial state variables [12,42]. At each moment, the resistances of the memristor elements M_1 and M_2 must be equal to each other.

Theoretically, the frequency of the generated output voltage signal is established with Equation (3.7) [12,42]:

$$f_0 = \frac{1}{2\pi R_{eqM1}C_1} \tag{3.7}$$

The supply voltage of the operational amplifier does not grow up instantly but increases almost linearly from 0 to 15 V for a very small time interval. That is the initial actuation pulse for creating the further signal generations in the Wien oscillator. The voltage signal generations could be actuated in several different ways, either by short external current or voltage impulse or by previously charging of some of the capacitors in the analyzed device. The initial normalized electric charge of a titanium dioxide memristor element is denoted with $a_M = q(t_0)/Q_d$, where $q(t_0)$ is the instantaneous electric charge accumulated in the memristor element, and Q_d is the maximum amount of charge accumulated in the memristor nanostructure [42]. The resistances of the memristor elements M_3 and M_4 must not be smaller than 21.1 kΩ. This requirement follows the stability criterion for the oscillator scheme [12,42].

3.2.2. Results and Discussion

The time graph of the generated output voltage signal for $a_{M3} = a_{M4} = 0.3$ is presented in Figure 84, especially for initial transient observation [19,42]. The duration of the initial transient in the oscillator is about 12 ms. After this process, the output voltage has amplitude and frequency stability. In this case, the normalized memristor charges of the memristor element M_1 and M_2 (a_{M1} and a_{M2}) are both 0.4 [12,19,42].

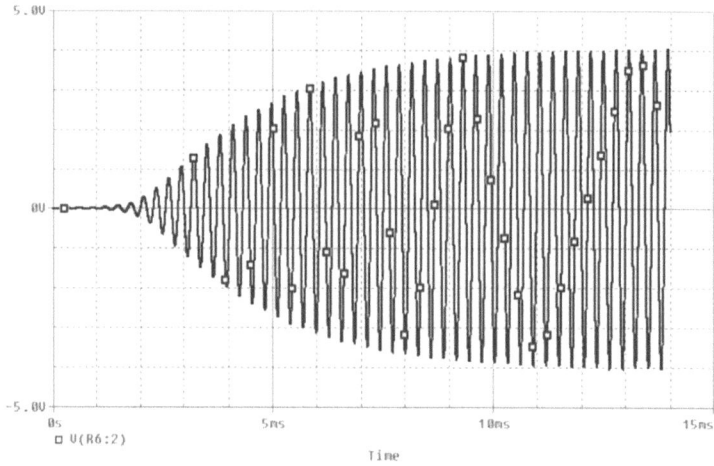

Figure 84. Time diagram of the output voltage of the memristor generator. In the beginning, a transient with a duration of about 12 ms is observed; the basic reason for this transient is the use of capacitors and resistors.

Theoretically, the maximum-to-minimum frequency ratio can be expressed as:

$$\frac{f_{max}}{f_{min}} = \frac{1}{2\pi R_{eqM1min}C_1} : \frac{1}{2\pi R_{eqM1max}C_1} =$$

$$= \frac{R_{eqM1max}}{R_{eqM1min}} = \frac{R_{OFF}}{R_{ON}} = \frac{16000}{100} = 160$$

(3.8)

This ratio is comparatively high and it could be concluded that a similar memristor-based Wien oscillator may almost cover the audio frequency range [12,42]. The phase portrait of the analyzed oscillator is shown in Figure 85 for visualization of the trajectory of the operating point in the field of the respective coordinates of the capacitor's voltage and current, which are most frequently used [12,19,42]. In this case, the phase portrait is obtained using the current–voltage characteristic of the capacitor C_2 during the operation of the generator circuit. The limit cycle derived for the established stable oscillations is visibly expressed [12,19,42].

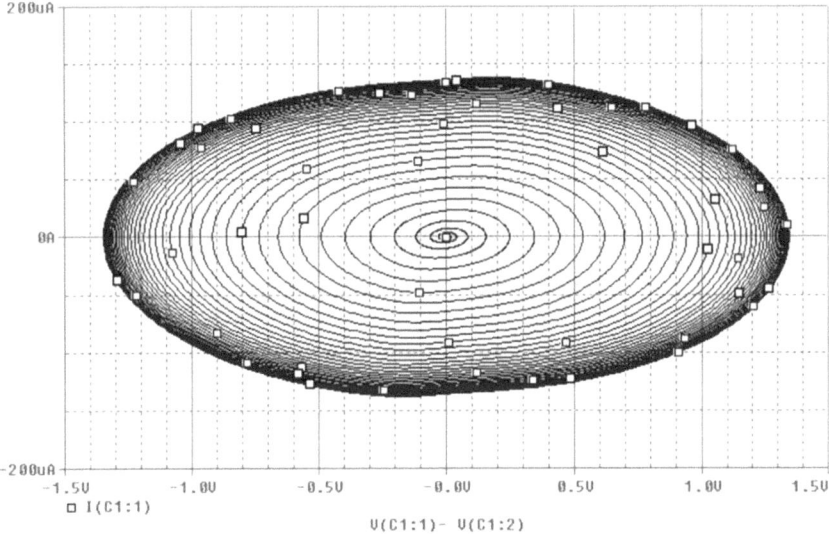

Figure 85. A phase portrait of the considered memristor-based generator, representing the trajectory of the memristor operating point in the field of the respective coordinates of the capacitor's voltage and current. The limit cycle, responsible for the derived stable oscillations, is clearly observable in the periphery of the curve.

From the results shown above, it is obvious that the memristor elements could be applied as trimming elements in oscillator devices. The memristor elements could replace some of the resistors in the Wien oscillator device. The memristor elements could be applied in the frequency-determining circuits as the Wien bridge and in the circuit for balancing the oscillator scheme. The ease of using memristor elements is the opportunity for their simple and fine adjustment by applying external pulses. In spite of the memristor's nonlinearity, it has been established that for a suitable mode, it does not introduce nonlinear distortions to the generated output signal. It has also been established that for very high frequencies, the memristor elements could operate with high levels of the applied signals without occurrence of breakthrough or another destructive effects. These phenomena could be explained with the high frequency of the generated signal and the inertia of the memristor components. The movement of the boundary between the doped and the undoped regions of the memristor element, in fact, has a very low amplitude, owing to the growing of the frequency, and the decreasing of corresponding flux linkage.

According to the state–flux characteristic of the investigated memristor, the state variable x is proportional to the flux linkage. This is the reason for determining very low divergence in the memristor state variable x, if the applied voltage has a high frequency [12,42]. The states of the memristor elements and their respective resistances are changing by applying pulses with large durations and low frequencies from external sources. By changing the memristor's resistance and electric charge, the frequency of the generated output signal could be tuned in a wide range. Memristor circuits similar to these shown above could be applied in audio-frequency signal generator devices [12,42].

3.3. Investigation of Anti-Parallel Circuit with Two Memristors

3.3.1. General Information and Mathematical Description of the Anti-Parallel Memristor Circuit

An anti-parallel memristor circuit with two memristor elements is investigated in an oscillator circuit [16,43]. The absence of detailed analysis of anti-parallel memristor circuits for the sine-wave regime is the basic motivation for the present research [13]. The modified linear drift memristor model proposed here is based on the GBCM model [14], but for simplicity, the model investigated is without activation thresholds. For the present investigations, an algorithm based on the finite difference method for numerical analysis is applied [16,43]. The main purposes of the present research are to acquire the basic important characteristics of the memristor circuit and its equivalent conductance for a sine-wave current source, and to derive the main results for hard-switching and soft-switching electric regimes [43].

The memristor circuit under analysis is presented in Figure 86 for derivation of its basic properties and behavior in electric fields [16,43]. The circuit contains two equivalent memristors with different initial values of the respective state variables x_1 and x_2. The anode of the first memristor element is connected to the cathode of the second memristor element, and the cathode of the first memristor element is connected to the anode of the second memristor element. The corresponding state differential equations of the memristor elements are as follows—System of Equations (3.9) [43]:

$$\frac{dx_1}{dt} = \eta_1 \, k \, i_1 = k \, i_1$$
$$\frac{dx_2}{dt} = \eta_2 \, k \, i_2 = -k \, i_2$$

$$(3.9)$$

where η_1 and η_2 are the respective polarity coefficients of the first and second memristor elements; for the first memristor element connected in a forward direction, the polarity coefficient is 1; for the second memristor element connected in a reverse direction, the polarity coefficient is equal to -1 [16,43]. According to the Kirchhoff's Current Law (KCL), the following relationship is expressed for the upper node (Figure 86) [12,16,43]—Equation (3.10):

$$j_e(t) = i_1 + i_2 \tag{3.10}$$

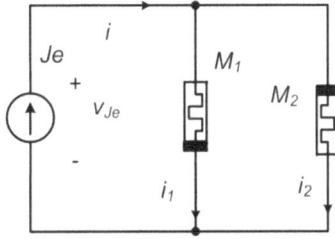

Figure 86. An anti-parallel circuit under analysis. The cathode of the first memristor element is connected to the anode of the second memristor element, and the anode of the first memristor element is connected to the cathode of the second memristor element. The applied current (unit: mA) can be described as: $j_e(t) = 0.1\sin\left(2\pi \times 40t - \frac{\pi}{3}\right)$.

By expressing the resistances of the memristor elements, the equivalent resistance R_{12} of the anti-parallel-connected memristor elements can be found [12,16,43]—System of Equations (3.11):

$$R_1 = (R_{ON} - R_{OFF})x_1 + R_{OFF}$$

$$R_2 = (R_{ON} - R_{OFF})x_2 + R_{OFF}$$

$$\Delta R = R_{ON} - R_{OFF} \tag{3.11}$$

$$R_{12} = \frac{R_1 R_2}{R_1 + R_2} = \frac{(R_{OFF} + \Delta R x_1)(R_{OFF} + \Delta R x_2)}{2R_{OFF} + \Delta R(x_1 + x_2)}$$

The voltage drop across the two memristor elements is expressed, using the Ohm's Law [12]—Equation (3.12):

$$v = j_e(t) \cdot R_{12} = j_e(t) \frac{(R_{OFF} + \Delta R x_1)(R_{OFF} + \Delta R x_2)}{2R_{OFF} + \Delta R(x_1 + x_2)} \tag{3.12}$$

The currents flowing through the two memristor elements are acquired using the current divider rule [12]—Equation (3.13):

$$i_1 = j_e(t)\frac{(R_{OFF}+\Delta Rx_2)}{2R_{OFF}+\Delta R(x_1+x_2)} \qquad i_2 = j_e(t)\frac{(R_{OFF}+\Delta Rx_1)}{2R_{OFF}+\Delta R(x_1+x_2)} \qquad (3.13)$$

A sinusoidal current source is applied for the computer simulation of the circuit [12,16,43]. Using Equations (3.10), (3.12) and (3.13), the KCL and the finite difference method, a pseudo-code-based algorithm is created and applied for analysis of the suggested memristor circuit [16,43].

The anti-parallel circuit presented in Figure 86 is analyzed for soft-switching and hard-switching modes [16,43] and its simulations were done in MATLAB [13]. For investigation of the anti-parallel memristor circuit for the soft-switching regime, a sinusoidal current source with the following current signal (unit: mA), described as $j_e(t) = 0.1\sin(2\pi \times 40t - \frac{\pi}{3})$, is applied.

3.3.2. Analytical Results. Investigation of the Anti-Parallel Circuit for the Soft-Switching Mode

The time graphs of the source current and the voltage drop across the anti-parallel-connected memristor elements, acquired after the numerical analysis, are shown in Figure 87 for visual illustration of the forms of the signals and their distortions. It is clear that the source current has a sine-wave shape, but the voltage across the memristors has a non-sinusoidal form, owing to the nonlinearity of the circuit [43].

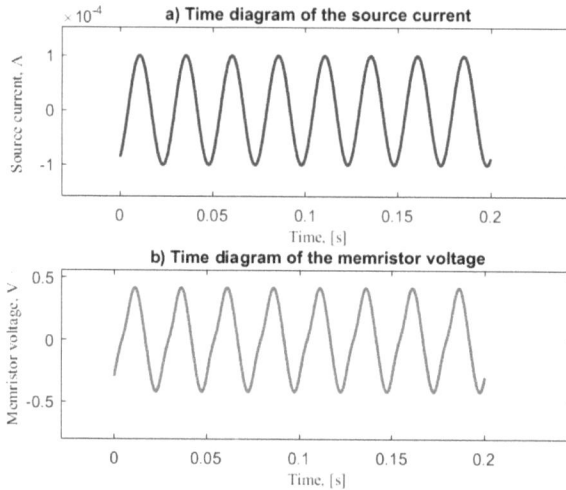

Figure 87. (a) Time graphs of the source current $j_e(t)$ and (b) time diagram of the voltage v across the memristor elements M_1 and M_2 for the soft-switching mode. The voltage drop across the memristors is a non-sinusoidal function.

The time graphs of the state quantities x_1 and x_2 for the two memristor elements are presented in Figure 88(a,b) for observation of their forms and performances in the time domain. It is obvious that the state quantities x_1 and x_2 do not attain their limiting values [16,43]. When the state variable x_1 grows up, the state variable x_2 decreases, due to the anti-parallel memristor biasing [16,43].

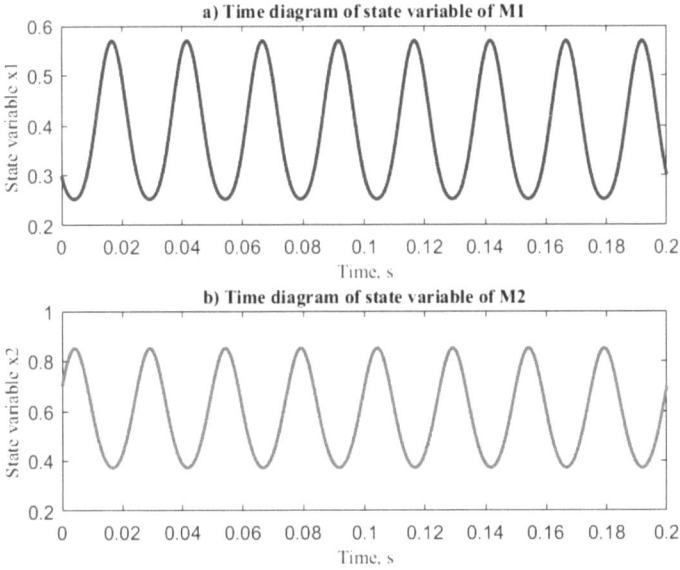

Figure 88. (a) Time diagrams of the state variable x_1 and (b) time diagram of the state variable x_2 of the memristor elements for the soft-switching regime, representing their change and behavior in the time domain.

The time graphs of the state variables presented in Figure 88(a,b) have non-sinusoidal shapes. The state–flux characteristics of the elements M_1 and M_2 are given in Figures 89 and 90, respectively. They are single-valued nonlinear functions. Owing to the anti-parallel bonding, they have different signs of their slopes, that is, the state–flux characteristic of the first memristor element is an increasing function while that of the second memristor element is a decreasing function [16]. When x_1 grows up, the state variable x_2 decreases, and vice versa. The applied flux linkage causes alterations in the state variables of the memristor elements in different directions.

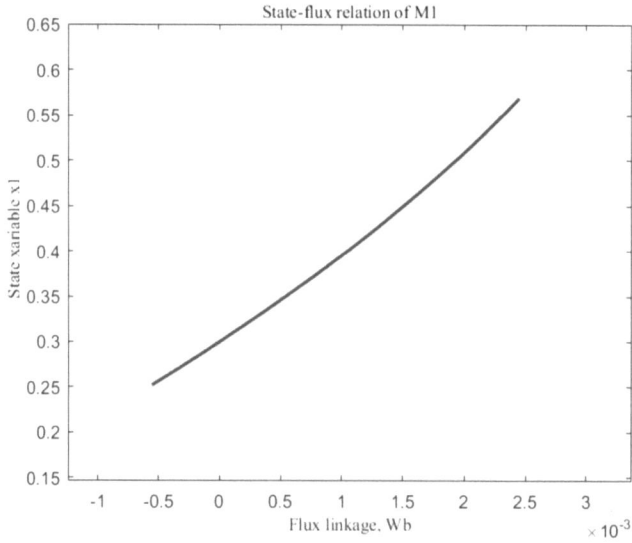

Figure 89. State–flux characteristic of the memristor element M_1 for the soft-switching mode; due to its forward biasing in the investigated circuit, containing anti-parallel-connected memristors, increasing the flux linkage Ψ causes the increasing of the state variable x_1.

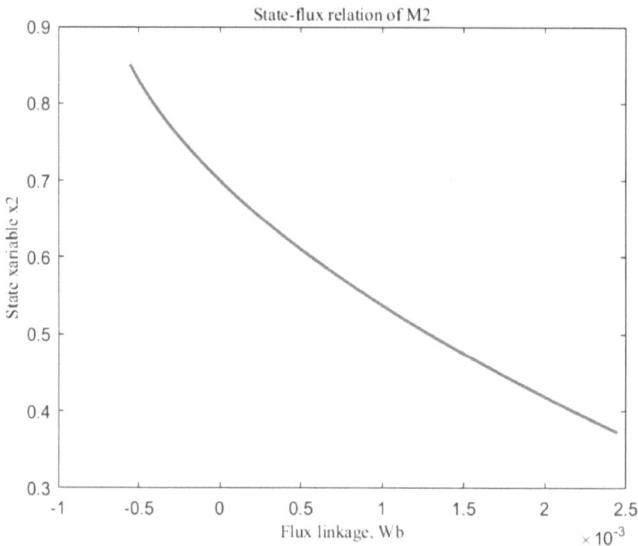

Figure 90. State–flux relationship of the second memristor M_2 for soft-switching state; due to the reverse biasing of the element in the circuit, increasing the flux linkage Ψ causes the decrease of the state variable x_2 of the memristor element.

The time graphs of the memristances of the elements in the investigated circuit are represented in Figure 91(a,b). Owing to the anti-parallel connection of the memristor elements, when the resistance R_1 grows up, then R_2 decreases, and vice versa.

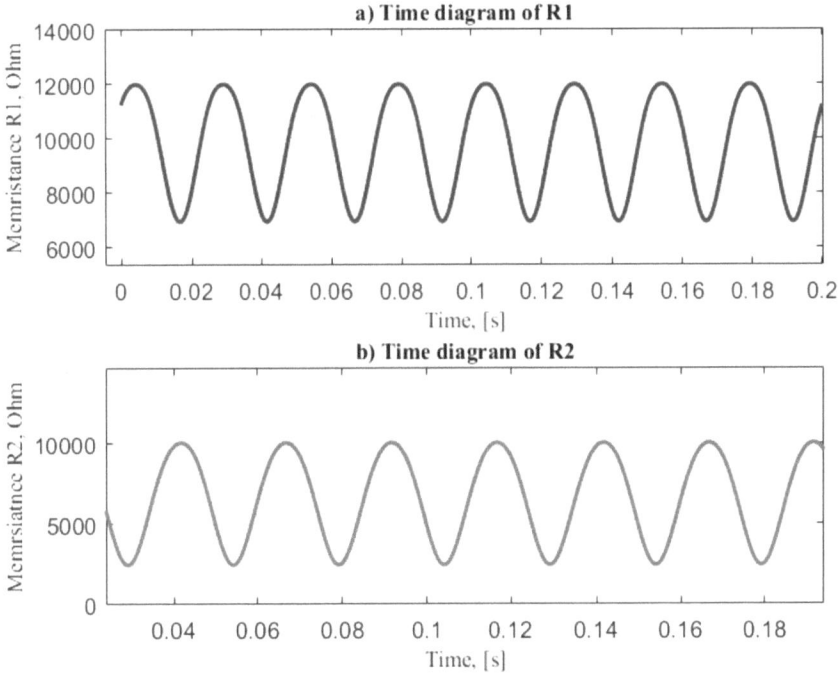

Figure 91. (a) Time graphs of the memristance of M_1 and (b) time diagram of the memristance of M_2 of the anti-parallel memristor circuit for the soft-switching regime, representing the change of the resistances of the elements M_1 and M_2 in different directions. The applied flux linkage Ψ causes the increase of the resistance of the first memristor element, and the decrease of the second one.

The time graph of the equivalent resistance R_{12} of the anti-parallel circuit is shown in Figure 92 for illustration of the unpredicted performance of the memristor circuit. It is clear that the time graph of the anti-parallel resistance of the memristor connection is a periodical, time-dependent and non-sinusoidal curve.

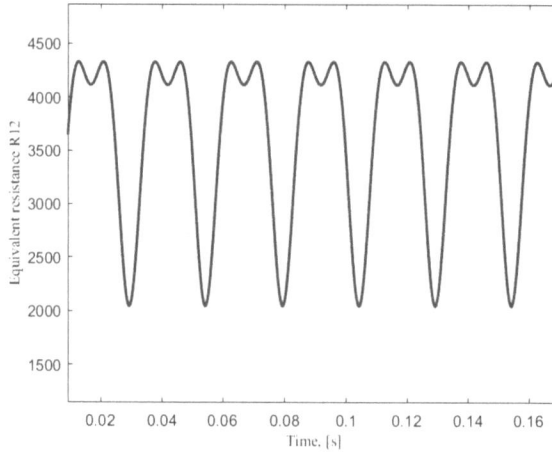

Figure 92. Time diagram of the equivalent resistance R_{12} of the anti-parallel memristor circuit for the soft-switching mode. The resistance changes in a comparatively narrower range.

The memristance–flux characteristic of the anti-parallel memristor circuit is shown in Figure 93. It is a single-valued function with a local maximum which is derived for a specific value of the resistances R_1 and R_2 according to System of Equations (3.11). The performance of the memristance–flux relationship of the anti-parallel memristor circuit is unpredicted, because it differs from the behavior of a single memristor, owing to the existence of its local maximum [16].

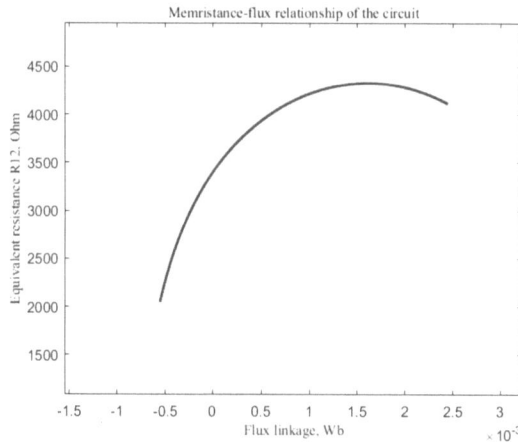

Figure 93. Memristance–flux characteristic of the anti-parallel memristor circuit for the soft-switching mode. The performance of the memristance–flux characteristic of the anti-parallel memristor connection is unpredicted because it differs from the behavior of a single memristor element due to the occurrence of its local maximum.

110

The respective i–v characteristics of the memristor elements M_1 and M_2 are shown in Figures 94 and 95. It is obvious that the current flowing through the second memristor M_2 is higher than the current through the memristor element M_1. This fact could be explained with the different initial values of the state variable and the different connection polarities of the memristor elements.

The current–voltage characteristic of the anti-parallel memristor circuit is shown in Figure 96. It is shown by a pinched hysteresis loop and it indicates that in this case, the parallel connection of two anti-parallel memristor components has a performance of a single memristor [16,43]. An interesting fact is the occurrence of a linear region in the current–voltage characteristic (Figure 96). This region corresponds to the OFF-resistance state of the circuit. The other nonlinear section of the current–voltage relationship is acquired when the circuit is in the ON-state with a lower resistance than that for the OFF-state.

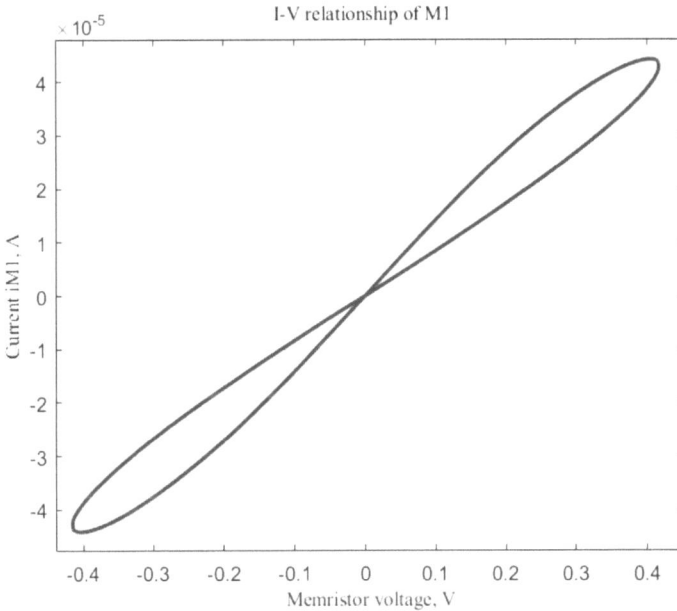

Figure 94. Current–voltage characteristic of the first memristor M_1 for the soft-switching mode.

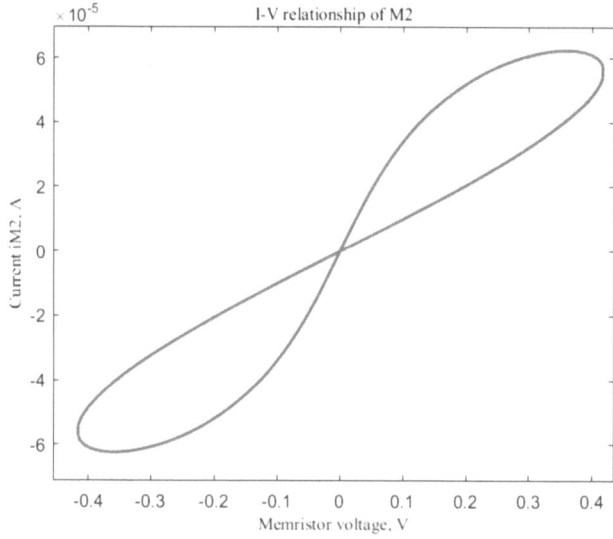

Figure 95. Current–voltage relationship of the first memristor element M_2 for the soft-switching mode. It is obvious that the area of this pinched hysteresis loop is larger than the that of the first memristor element M_1; this fact could be explained with the different initial values of the memristor state variables x_1 and x_2 and the different biasing polarity of the memristor elements.

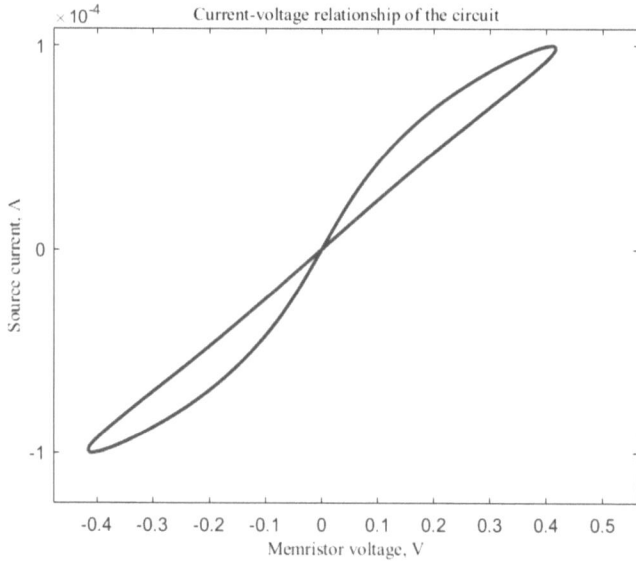

Figure 96. Current–voltage characteristic of the whole anti-parallel memristor circuit for the soft-switching regime.

3.3.3. Analysis for the Hard-Switching Mode

For investigation of the anti-parallel circuit for the hard-switching mode, a sinusoidal current source with the following signal (unit: mA), described as $j_e(t) = 1 \times \sin(2\pi \times 40t - \frac{\pi}{3})$, is applied. The time graphs of the source current and the voltage across the memristors are shown in Figure 97(a,b). It is observable that the source current has a sine-wave form, but the voltage across the passive elements has a strongly non-sinusoidal shape by reason of the extended nonlinearity of the circuit for the hard-switching regime. The time graphs of the state variables x_1 and x_2 for the hard-switching mode are given in Figure 98(a,b). It is obvious that the state variables attain their limiting values—zero and unity, that is, if x_1 has a value of 1, then the state variable x_2 is equal to 0, and vice versa. The state–flux characteristics of the two memristor elements are shown in Figure 99(a,b). It is obvious that they are multi-valued hysteresis functions.

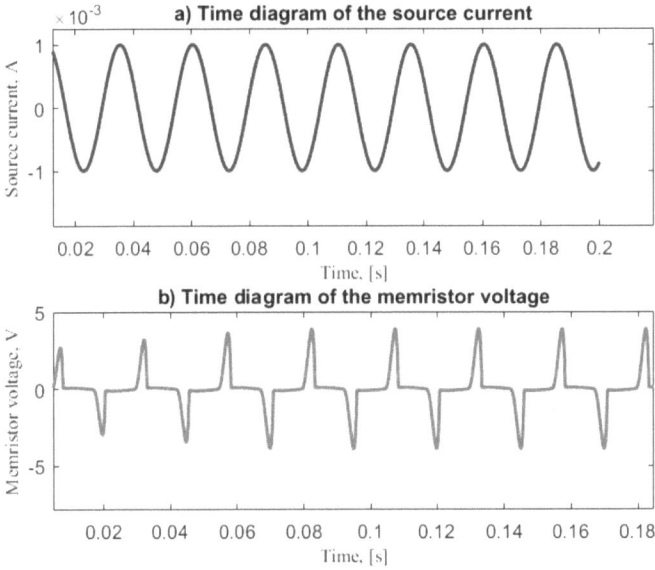

Figure 97. (a) Time graphs of the source current $j_e(t)$ and (b) time diagram of the voltage v_j across the memristors M_1 and M_2 for the hard-switching mode.

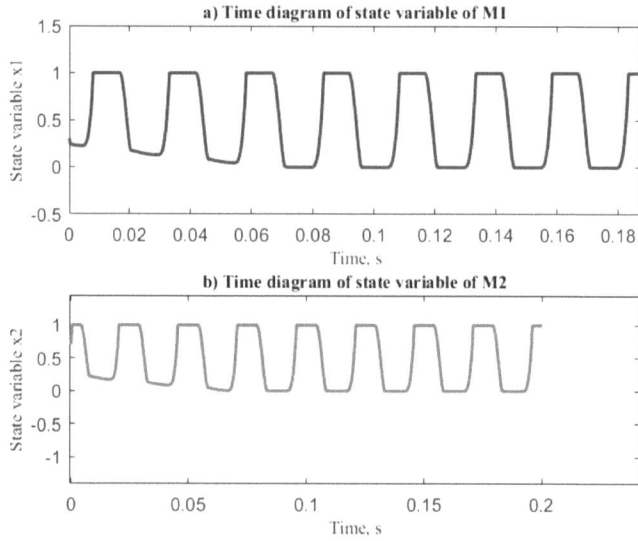

Figure 98. (a) Time graphs of the state variable x_1 and (b) time diagram of the state variable x_2 of the memristor elements for the hard-switching mode. The state variables attain their limiting values—zero and unity. If x_1 has a value of 1, then x_2 is equal to 0, and vice versa; the memristors are switched in the anti-phase regime.

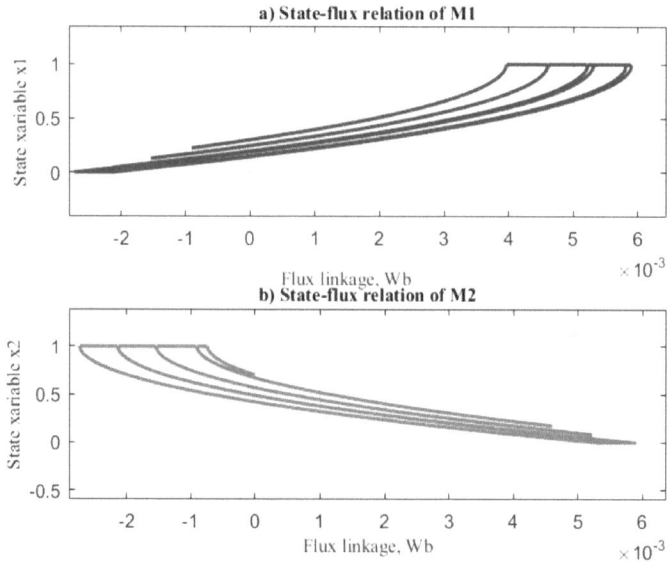

Figure 99. (a) State–flux relationships the memristor element M_1 and (b) state-flux relationship of M_2 for the hard-switching mode. The applied flux linkage Ψ causes the increase of the state variable of the first memristor element and the decrease of the second memristor element.

114

Owing to the different biasing directions of the memristors, the slope and the corresponding first derivative of the state–flux relationships are different for the two memristor elements. When x_1 grows up to unity, then the state variable x_2 decreases to zero, and vice versa.

The memristance–flux relationship of the anti-parallel memristor connection is given in Figure 100. The state–flux characteristic is a multi-valued hysteresis function with local maximums. The memristances of the two memristor elements are reversely proportional to their state variables.

The time graphs of the memristances of the two memristor elements are given in Figure 101(a,b). It is obvious that their resistances attain their limiting values—100 Ohms and 16 000 Ohms, respectively, without the transition time intervals between the ON- and OFF-states [16,43]. In the time intervals, when the elements work in a soft-switching state, the total resistance of the scheme is higher than 100 Ohms [16,43]. The time graph of the total resistance of the anti-parallel connection is given in Figure 102. It is clear that for more time intervals, the first or the second element operates in a hard-switching mode and has its maximal conductance [16].

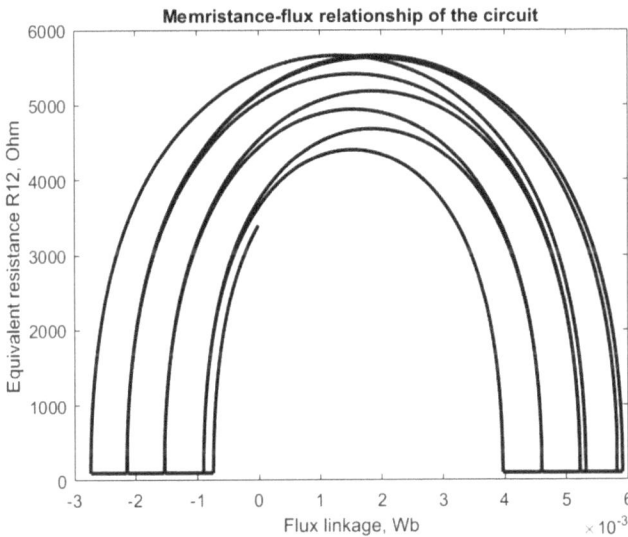

Figure 100. Memristance–flux characteristic of the complete anti-parallel memristor circuit for the hard-switching mode.

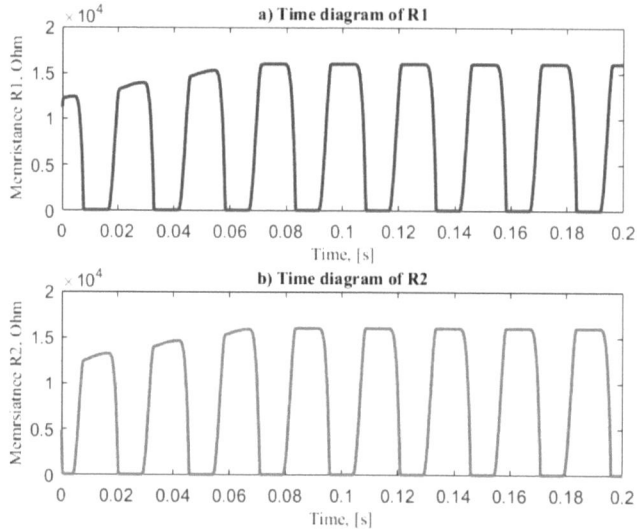

Figure 101. (a) Time graphs of the memristance of M_1 and (b) time diagram of the memristance of M_2 of the anti-parallel circuit, containing the memristors M_1 and M_2 for the hard-switching regime.

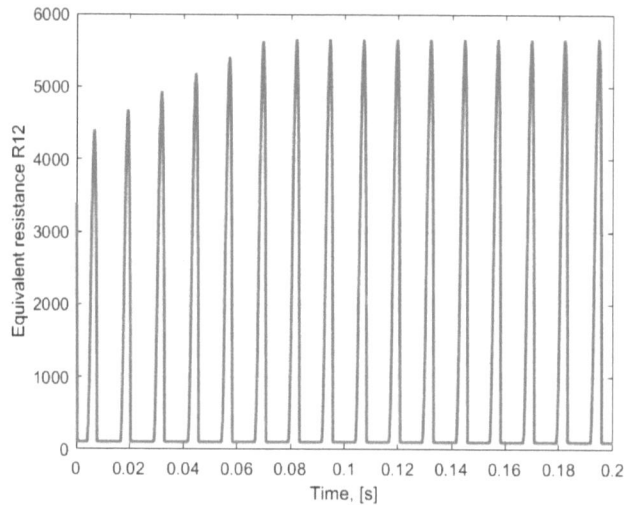

Figure 102. Graph of the total resistance of the anti-parallel circuit R_{12} for the hard-switching regime. In the higher duration of the investigation, the total resistance of the anti-parallel memristor circuit is close to 100 Ohms.

By virtue of the mentioned information, the total resistance of the circuit for the long-time intervals has its minimum value, which is lesser than the ON-resistance of a single element, i.e., 100 Ohms [16,43].

A remarkable fact could be derived if the switching speed of the memristors is very high and the switching process from the OFF-state to the ON-state is established almost instantly. Then, the resistance of the whole circuit would be a constant, because one of the memristors is always in the ON-state. Practically, a very short time interval is needed for switching the single memristor from the OFF-state to the ON-state, and the described mode is only theoretically realized.

The current–voltage characteristics of the particular elements are shown in Figure 103(a,b). These functions in the current case are anti-symmetrical. For the hard-switching regime, the memristors have a rectifying effect and their performance is analogous to that of the semiconductor diodes [16]. When the first memristor element is in a fully open state, the second memristor element reaches its fully closed position, and vice versa. The current–voltage relationships of the equivalent anti-parallel circuit are presented in Figure 104. Due to the anti-parallel biasing of the memristor elements, the equivalent memristor scheme has a symmetrical pinched multi-valued current–voltage relationship and does not have rectifying properties in this situation.

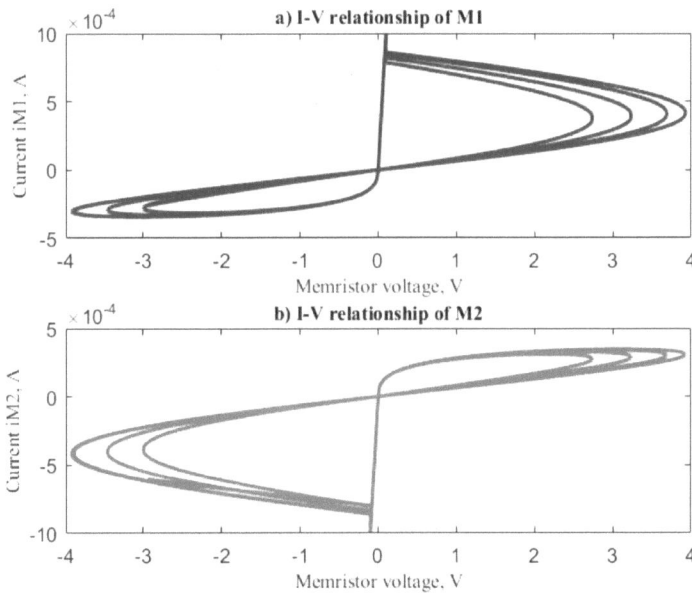

Figure 103. (a) Current–voltage characteristics of the first element M_1 and (b) current-voltage characteristic of the second memristor M_2 for the hard-switching mode. These graphs in the present case are anti-symmetrical; for the hard-switching regime, the memristors have a rectifying effect and their performance is similar to that of the rectifier diodes.

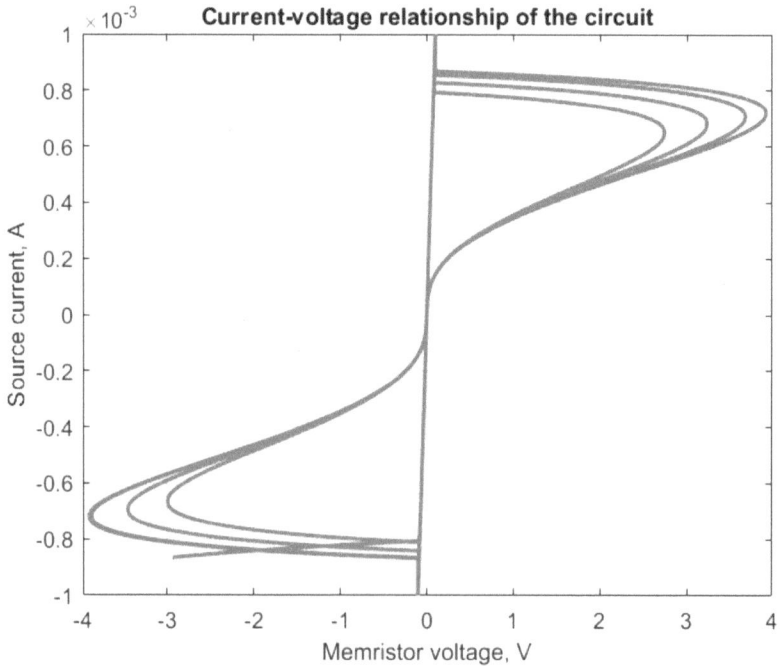

Figure 104. Current–voltage characteristic of the complete anti-parallel memristor scheme for the hard-switching regime.

Owing to the high nonlinearity of the memristor elements for the hard-switching mode, the corresponding memristance–flux characteristics and the current–voltage relationships are multi-valued functions. The maximum values of the total memristance of the anti-parallel circuit are in a range from 4400 Ohms to 5600 Ohms. When the equivalent resistance of the anti-parallel circuit under analysis has its maximal values, then both the elements are operating in a regime close to a fully closed state [16,43].

When the total resistance of the analyzed anti-parallel circuit is smaller than 100 Ohms, then at least one of the memristor elements is working in a fully closed state, of which the corresponding state variable is equal to unity. The corresponding resistance of the circuit is close to 100 Ohms [16,43].

3.3.4. Discussion of Results

After finalizing the investigations and the computer analysis of the anti-parallel memristor circuit, several conclusions could be made. The analytical results verify the theoretical analysis made in Section 3.3.1. The influence of the source has a different effect on the two memristor elements due to their different biasing.

For a sine-wave source current signal with amplitude of 0.1 mA, both the elements operate in a soft-switching regime. For the soft-switching state, if one of the memristors has a growing conductance, the resistance of the other memristor element increases [16,43]. The whole memristor circuit has a single-valued pinched hysteresis function of the current–voltage characteristic. In the current situation, it behaves as a single memristor element.

For a sine-wave signal with an amplitude of 1 mA, both the memristors operate in a hard-switching state. In many time intervals with a reasonably long duration, the first memristor element is in a fully open state, while the second memristor element is in a totally closed state, and vice versa. Then, the anti-parallel circuit has its minimum resistance lower than 100 Ohms.

During the other time both the memristor elements operate in an active state, Therefore, the state variables are in the interval (0, 1). The corresponding current–voltage characteristic of the whole anti-parallel circuit is shown by a symmetrical multi-valued graph, despite the fact that each memristor element in the anti-parallel circuit has an anti-symmetrical current–voltage characteristic. Then, the corresponding behavior of the anti-parallel memristor scheme is different from that of the known rectifier components. For the hard-switching regime, the circuit has a higher nonlinearity [16].

3.4. Integrator Scheme with a Memristor

3.4.1. General Information for the Analyzed Circuit and the Applied Memristor Model

Integrator schemes are important and extensively applied modules in many complex radio electronic circuits. The awareness of their new schematic design is basically related to their universal applications in electronics. The goal of this analysis is to offer an inclusive investigation of a suggested memristor-based integrator device with an operational amplifier by the author. The scheme under analysis is based on the traditional resistor–capacitor integrator device with an operational amplifier. In the proposed circuit, the resistor liable for the integrating processes is replaced by a memristor [36].

Integrator schemes are main modules of large electronic circuits. They are applied for acquiring a signal, proportional to the time integral of the input signal. If the input signal is a sequence of rectangular impulses with different directions, the corresponding output signal is a piecewise linear function with growing and decreasing regions [36]. The integrator units are applied in radio-electronics, automatics and many other areas of technical industry [36].

Many measurements of electronic modules also enclose integrator circuits. The conventional integrator devices contain resistors, capacitors and operational amplifiers [36]. To the best of the author's knowledge, derived after the reference check, there is definitely an absence of complete results derived by measurements or by investigations of memristor integrators with the well-identified memristor models [5–7,33]. The motivation for the present analysis is to fill this absence, offering a detailed investigation of a memristor–capacitor integrator device with an operational amplifier [14]. For the present investigation, an altered strongly nonlinear memristor model suggested by the author in Reference [33] is used. Several basic memristor models are applied as well [5–7,33]. A comparison with a traditional Resistor-Capacitor (RC) integrator device with an operational amplifier is accomplished [36]. The ability of the used memristor model [33] for operation in multipart electronic devices containing integrator units is established [14].

The suggested altered Biolek model [6] applied here is described by System of Equations (3.14) [6,33]. If both the coefficient m and the sensitivity threshold v_{thr} of the memristor element are zero, then the suggested improved Biolek model is modified in the form of the standard Biolek model [33].

$$\frac{dx}{dt} = k\eta\,i\left[\frac{-(x-1)^{2p}+m\left(\sin^2(\pi x)\right)+1}{1+m}\right], \qquad v(t) \leq -v_{thr}$$

$$\frac{dx}{dt} = k\eta\,i\left[\frac{-x^{2p}+m\left(\sin^2(\pi x)\right)+1}{1+m}\right], \qquad v(t) > v_{thr} \tag{3.14}$$

$$\frac{dx}{dt} = 0, \qquad -v_{thr} < v(t) \leq v_{thr}$$

$$v = R\,i = \left[(R_{ON}-R_{OFF})x + R_{OFF}\right]i$$

The flux–charge and the current–voltage characteristics of the memristor element, according to the reference Pickett memristor model [7], are acquired in PSpice environment [19] for a sinusoidal voltage signal, and they are applied for tuning the used altered memristor model [33]. The corresponding flux–charge and current–voltage relationships of the suggested modified memristor model [33] are derived after a number of simulations and adjustment of the improved memristor model, in accordance to the reference Pickett model [6]. These characteristics are established by simulations in MATLAB [13]. After finishing the tuning procedures and deriving a reasonably good matching between the main characteristics—the current–voltage and flux–charge relationships, the improved memristor model offered by the author in [33] is applied for analysis of the memristor-based integrator device [36].

A resistor–capacitor integrator scheme with an operational amplifier [12] and its corresponding memristor analogue are shown in Figures 105 and 106. The resistor R_2 is applied for incomplete discharging of the capacitor during the operation of the integrator circuit. The analyses are made in MATLAB environment [13].

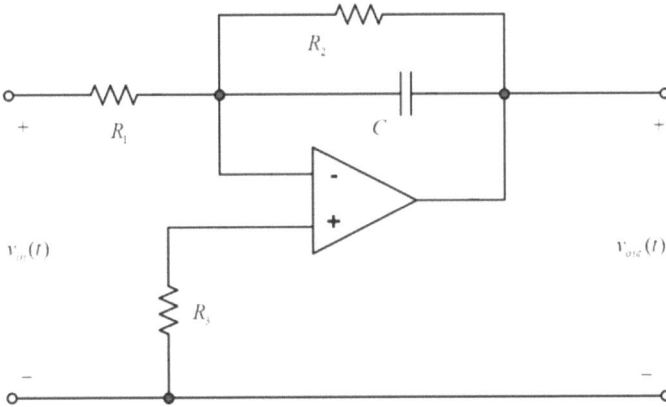

Figure 105. Classical Resistor-Capacitor (RC) integrator device with an operational amplifier, representing the basic integrator structure. It can be used as a basis for construction of the new memristor integrator.

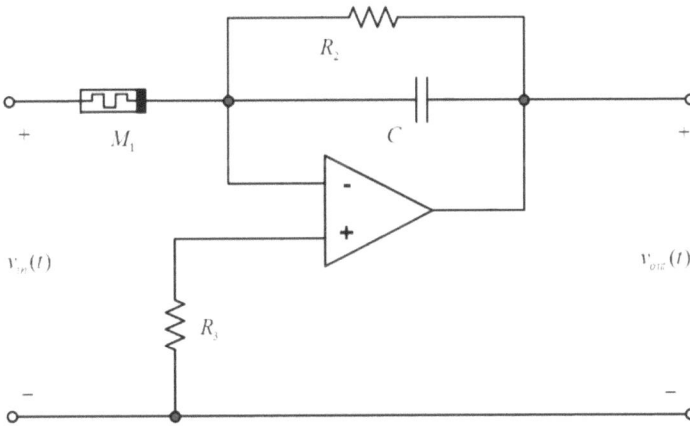

Figure 106. An integrator device with an operational amplifier, a capacitor and a memristor, based on the traditional structure, shown in Figure 105. The resistor R_1 is replaced with the memristor element M_1.

3.4.2. Results and Discussion

The graphs of the corresponding input and output voltage signals are presented in Figure 107(a,b) to be compared with the results acquired by the BCM memristor model. In both schemes, the capacitor has a capacitance of 20 µF. The resistor

R_1 applied in the first scheme has a resistance of 7 kΩ. This resistance is close to the average value of the resistance of the memristor element in the time domain. It is obvious that the magnitude of the output voltage signal for the memristor integrator is higher than that of its resistive analogue. This is a benefit of the suggested memristor-based integrator by the author [36]. It is observable from the graphs presented in Figure 107(a,b) that the output voltages of both the traditional and the memristor integrators are proportional to the time integral of the input voltage signals. Owing to the use of the inverting input of the operational amplifier, the output voltage signal changes in the inverse direction compared to the input voltage [12]. After an additional analysis for a very long-time interval, it is established that due to the influence of the capacitor, there is a transient in the integrator device. In this time interval, the magnitude of the output voltage signal is growing up with the time to a certain steady value [36]. It is established that the transient in the suggested memristor-based integrator is slightly shorter than the transient of the traditional scheme [12].

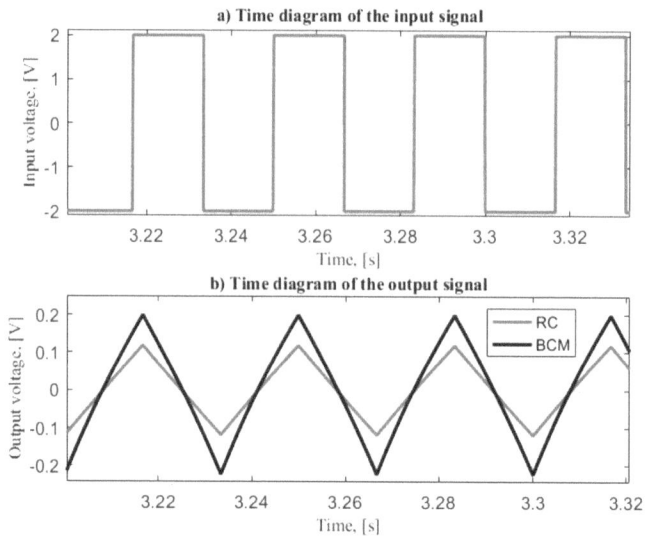

Figure 107. (a) Time diagram of the input voltage signal and (b) time diagram of the corresponding output voltage for the resistor–capacitor integrator device with an operational amplifier, and for the memristor-based integrator device, using the BCM memristor model.

A comprehensive time graph of the input and the output voltage signals according to several memristor models [5–7,36] is shown in Figure 108. The graphs of the output voltage derived by the application of the BCM, Biolek and the altered Biolek models cited above are very close to each other. For all the characteristics derived by the memristor-based device, the amplitude of the output voltage signal is higher than the amplitude of the output signal of the traditional integrator scheme [12].

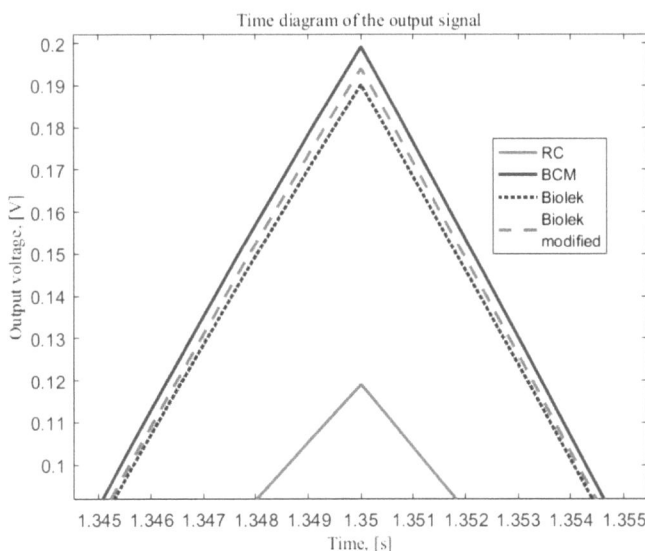

Figure 108. Complete time graphs of the output voltage signal for the RC integrator device with an operational amplifier and the memristor-based integrator scheme, according to the BCM model, Biolek model and the altered Biolek model. The respective time graphs for these models are very close to each other.

It is established that the applied altered memristor model is able to illustrate the behavior of the memristor-based integrator electronic device [33,36]. The time graphs of the resistance of the memristor element in the operation processes of the integrator scheme for the BCM model, Biolek model and the suggested altered Biolek model are shown in Figure 109, and they are almost identical [36].

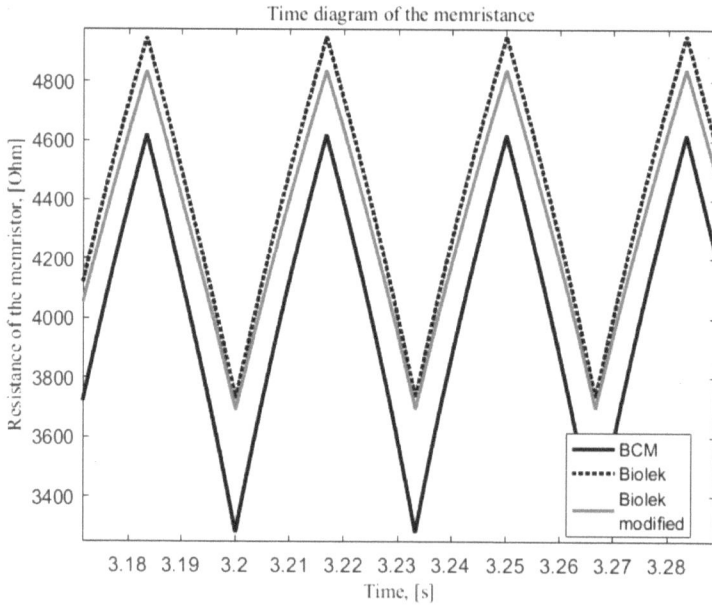

Figure 109. Comprehensive time diagrams of the memristance M according to several main memristor models (BCM, and Biolek) and the improved Biolek model. The derived time diagrams for these three models are very close one to another.

It is obvious that the resistance of the memristor M changes in the range between 3300 Ω and 5000 Ω. The range of the memristance is almost the same for all the applied models [5–7,33]. The maximum possible value of the memristance is 16 kΩ, so it could be established that in the present case, the memristor works in a soft-switching state. The memristance–flux characteristics of the memristor for the BCM model, Biolek model the applied memristor models are shown in Figure 110(a,b). They all are approximately monotonically increasing single-valued functions owing to the memristor element operation in a soft-switching regime [7,33]. The acquired state–flux characteristics for these are identical [36]. The corresponding current–voltage characteristics of the memristor for these three models are given in Figure 111(a–c). It is established that they approximately match each another. According to the performance of the current–voltage relations acquired by the suggested altered Biolek models, the memristor operates in a soft-switching regime. The behavior of the suggested altered Biolek model is similar to the performance of BCM and Biolek memristor models [5–7,33].

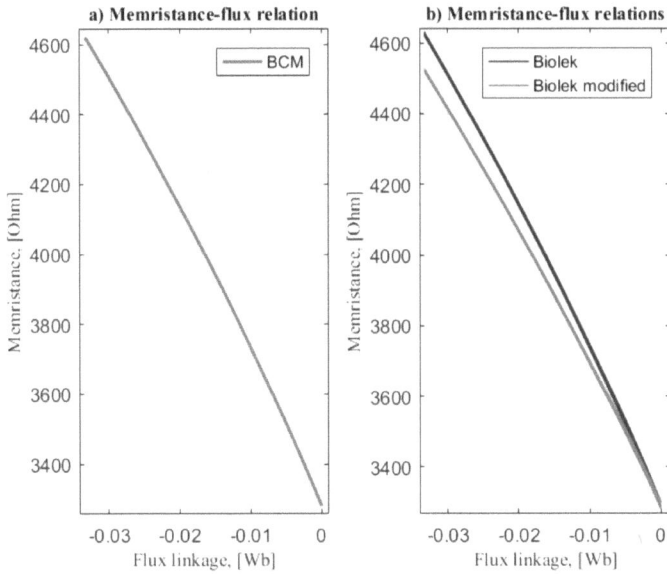

Figure 110. (a) Memristance–flux characteristics for the BCM model, (b) Memristance-flux relationships of Biolek memristor model and of the suggested improved Biolek model, which are very close to each other.

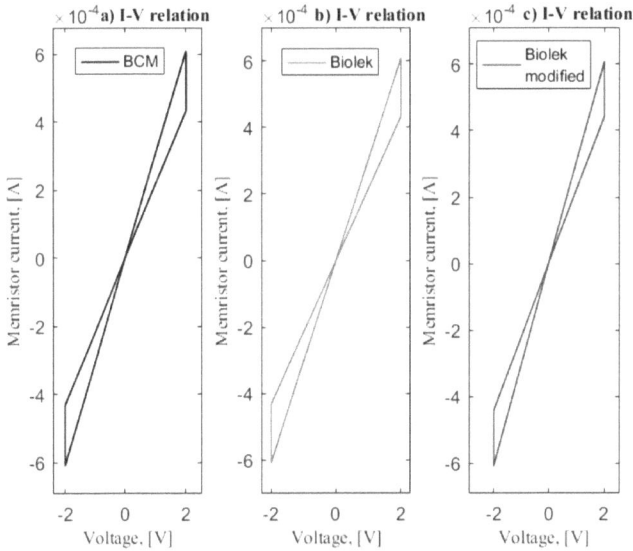

Figure 111. (a) Current-voltage characteristics of the memristor according to BCM model; (b) Current-voltage relation according to Biolek model; (c) Current-voltage characteristics according to the modified Biolek model.

After many computer analyses, it has been established that if the resistor R_1 in the RC integrator scheme with an operational amplifier is replaced by a memristor element, the scheme is still able to work as an integrator module [36].

Benefits of the offered memristor–capacitor integrator device with an operational amplifier are the lower duration of the transient and the higher magnitude of the output voltage signal [36]. By virtue of the nano-scale sizes of the memristor element, the suggested integrator device could be realized in integrated electronic schemes with very high concentrations. The integration procedure of rectangular voltage pulses is a linear one and it could be done by the application of a linear RC scheme with an operational amplifier. Though the memristor element is a nonlinear one, after replacement of one of the resistors with a memristor, the device operates as an integrator scheme. It is established that the transient in the memristor integrator module is shorter than that of the traditional integrator. The magnitude of the output voltage signal for the memristor integrator is higher than the amplitude of the output signal of the traditional integrator. These details are benefits of the suggested memristor integrator scheme [36]. In the operation mode of the proposed scheme, the memristor element works in a soft-switching regime. The computer analyses are realized by applying several main memristor models and the suggested in a different research improved memristor model by the author in different research areas [33,36]. The corresponding graphical results for these models are similar to each other. Then, it could be established that the altered Biolek model suggested by the author is suitable for analysis of memristor-based electronic devices containing integrator modules, operating in an impulse state [36]. Owing to the nano-scale sizes of the memristor elements, the suggested memristor integrator module could be incorporated into very-large-scale integrated circuits.

CHAPTER IV

Analysis of Memristor Networks

4.1. A Memristor Perceptron for Logical Function Emulation

4.1.1. General Information and Mathematical Description of the Investigated Memristor-Based Perceptron

The artificial neural networks are very important modules in almost all the present electronic and computer devices and systems [44]. They could represent the behavior and operation of biological neural systems and are used for solving many technical issues, such as signal processing, data clustering, image recognition, and decision-making [44]. The perceptron is a simplest kind of neural network. It is a nonlinear neuron and could implement the basic learning processes. It could also emulate several basic logical functions [44]. For the classical artificial neurons, Complementary-Metal–Oxide–Semiconductor (CMOS) circuits are used in the synaptic-weight circuits [44].

In the last few years, many scientists and engineers have paid attention to memristor neural networks because memristor elements and circuits have a memory effect and are appropriate for emulating the biological synapses by changing and holding their state [3,4]. Several different memristor synapses, such as bridge synapse and simple one-memristor synapse for neurons, are investigated [4,45]. One of the main advantages of the memristor-based synapses is the nano-scale dimensions of the memristor element and the possibility for high-density integration in crossbar-like structures together with CMOS elements [5–7,14]. The lack of detailed analysis of memristor-based linear synapses with two serial-biased memristors is a motivating factor for the present investigations. The basic purpose of the present research is to realize a complete investigation of the proposed memristor synapse by the author. The perceptron circuit for emulating the logical functions "OR" and "AND" is shown in Figure 112 [44] for describing the network structure and signals.

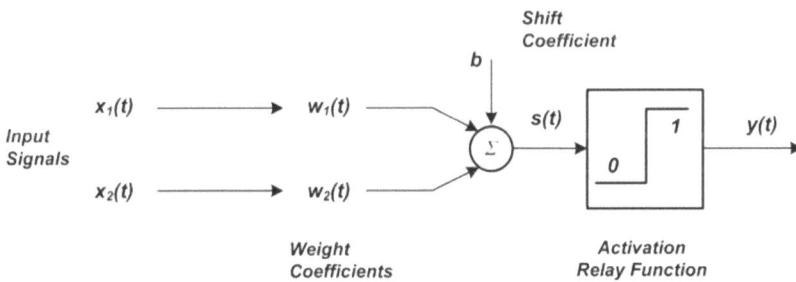

Figure 112. A perceptron circuit representing the basic structure of the considered artificial neuron and the used signals.

The simplest memristor synapse with one memristor [4] is presented in Figure 113(a). Its basic advantage is the use of only one memristor per synapse. The main disadvantages are the dependence of its operation on the input resistance of the summing device and realization of only positive weights in the interval (0, 1). The proposed memristor-based synapse by the author in the present research is presented in Figure 113(b) [46] for representation of the input–output relationships. The basic advantage of this memristor synapse is the simplified tuning of the weight and the lesser dependence of its operation on the input resistance of the next module—the summing device. The main disadvantage is the use of two memristors per synapse. Its synaptic weight could be changed in the interval (0, 1).

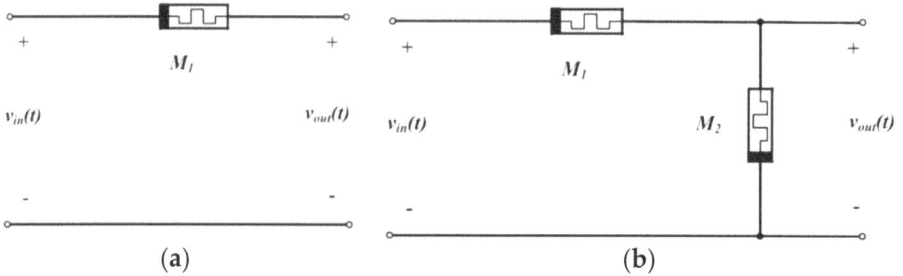

Figure 113. (a) A simple synapse with one memristor element; (b) a memristor-based anti-serial synapse circuit with two memristors, connected in an anti-series connection

The sampled time variable is denoted by t. The input binary signals $x_1(t)$ and $x_2(t)$ are first multiplied by the initial values of the respective synaptic weights $w_1(t)$ and $w_2(t)$. The weight coefficients are equal to the transfer function of a circuit containing memristors. There is a possibility for updating the synaptic weights by changing the memristor state variables x_1 and x_2. The weighted signals are applied to a summing device with a shift coefficient b. The output signal of the summing device $s(t)$ is [44,46] Equation (4.1):

$$s(t) = w_1(t)x_1(t) + w_2(t)x_2(t) + b \tag{4.1}$$

The acquired signal $s(t)$ acts as an input to a module with a relay activation function and with a threshold of $\Theta = 0$. The output signal $y(t)$ of the relay element is presented with Equation (4.2) [44,46]:

$$y(t) = stp[s(t)] = \begin{cases} 0, & s(t) < \Theta \\ 1, & s(t) \geq \Theta \end{cases} \tag{4.2}$$

The error signal $e(t)$ is derived by applying the present value of the output signal $y(t)$ and the desired (target) output signal $d(t)$, which is evaluated with respect to the logical functions "AND" and "OR" [44].

The error signal $e(t)$ is expressed as a difference between the desired (target) signal $d(t)$ and the output signal $y(t)$ [44]—Equation (4.3):

$$e(t) = d(t) - y(t) \tag{4.3}$$

The signal of the error $e(t)$ is used to be multiplied with the input signals $x_1(t)$ and $x_2(t)$, respectively, for realizing the correcting tuning amounts Δw_1 and Δw_2 and Δb of the synaptic weights w_1 and w_2 and of the shift coefficient b, which are written with the next Equation (4.4) [44,46]:

$$\Delta w_1(t) = x_1(t)e(t)$$
$$\Delta w_2(t) = x_2(t)e(t) \tag{4.4}$$
$$\Delta b(t) = e(t)$$

The new values of the synaptic weights w_{1new}, w_{2new}, and b_{new} are acquired by summing their old values w_{1old}, w_{2old}, and b_{old} with the correcting amounts Δw_1, Δw_2, and Δb. After updating the synaptic weights, after one epoch, all the input data are again applied to the circuit. The new values of the synaptic weights are expressed by the following Equation (4.5) [44,46]:

$$w_{1new}(t + \Delta t) = \Delta w_1(t) + w_1(t)_{old}$$
$$w_{2new}(t + \Delta t) = \Delta w_2(t) + w_2(t)_{old} \tag{4.5}$$
$$b_{new}(t + \Delta t) = \Delta b(t) + b_{old}(t)$$

where Δt is the time interval between two neighboring epochs. This time interval is used for updating the synaptic weights.

4.1.2. Results and Discussion

The preliminary assessments of the synaptic weights and the shift coefficient are chosen in accordance to the possibility for isolating the logical unity and zero for the logical functions "OR" and "AND". The initial sampled data are applied to the inputs of the perceptron scheme. The results, presented in Figures 114 and 115, confirm that the memristor-based perceptron effectively emulates the logical functions "OR" and "AND" by learning and training for 4 epochs [46]. The final 5th epoch is additional and is applied to verify that after adjusting the synaptic weights, they remain constant quantities if another logic sequence is applied to the inputs of the perceptron. After the final epoch, the error signal of the neuron $e(t)$ is zero. The weights tuning amounts of the synaptic weight Δw and the shift coefficient Δb have their values of -1 and 1, respectively. Applying the final values of the weights of the adjusted perceptron, the separated line equation for the logical functions is derived—Equation (4.6) [44,46]:

$$w_{1final}x_1 + w_{2final}x_2 + b_{final} - \Theta = 0 \qquad (4.6)$$

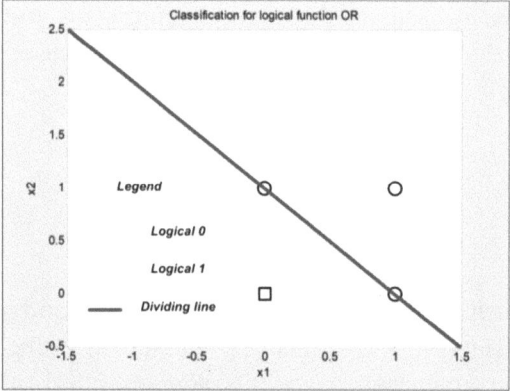

Figure 114. Representation of the results and the memristor perceptron dividing line for emulation of the logical function "OR", after finishing the learning processes.

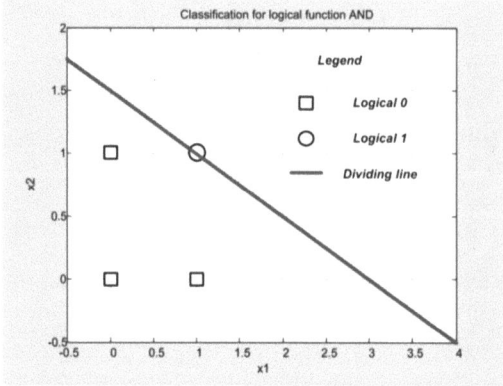

Figure 115. Results after emulation of the logical function "AND", and the derived separating line of the memristor perceptron.

The mathematical expression of the state–flux characteristic of the memristor elements, used in the synaptic circuits (Figure 112(b)), is shown in the next Equation (4.7) [4]:

$$x = x_0 + \eta \frac{\mu R_{ON}}{D^2} \int_{t_0}^{t} i \, dt' \qquad (4.7)$$

132

The variable x_0 is the initial state variable of the memristor element [4]. At the initial moment, the first memristor M_1 is set to the OFF-state and $x_{01} = 0$, and the corresponding polarity coefficient η is 1 because the memristor element is forward-biased [14].

The second memristor element M_2 is put in a fully closed state and the corresponding state variable x_{02} is 1. Due to its reverse-biasing with respect to the first memristor, its polarity coefficient η_2 is -1. The resistance of the first memristor element M_1 is expressed by Equation (4.8) [14,46]:

$$M_1 \approx R_{OFF}(1 - x_1) = R_{OFF}\left(1 - \frac{\mu R_{ON}}{D^2}\int_{t_0}^{t} i\, dt'\right) \tag{4.8}$$

The resistance of the second memristor element M_2 in the synapse is expressed by Equation (4.9) [14,46]:

$$M_2 \approx R_{OFF}(1 - x_2) = R_{OFF}\frac{\mu R_{ON}}{D^2}\int_{t_0}^{t} i\, dt' \tag{4.9}$$

The quantity R_{OFF} depicts the resistance of the memristor element in the OFF-state [4]. According to Kirchhoff's Voltage Law, by applying Equations (4.8) and (4.9), the total resistance of the memristor synaptic circuit R_{eq} is acquired—Equation (4.10) [46]:

$$R_{eq} = M = M_1 + M_2 = R_{OFF} \tag{4.10}$$

The synaptic weight (transfer function) w of the memristor-based synaptic circuit shown in Figure 113 is given in Equation (4.11) [4,14,46]:

$$w = \frac{M_2}{M_1 + M_2} = \frac{\mu R_{ON}}{D^2}\int_{t_0}^{t} i\, dt' = k\int_{t_0}^{t} i\, dt' \tag{4.11}$$

Using the Ohm's Law for the same circuit, supplied by a voltage input signal, a similar relationship is obtained—Equation (4.12) [46]:

$$w = k\int_{t_0}^{t} i\, dt' = k\int_{t_0}^{t} \frac{v_{in}}{R_{OFF}}dt' = \frac{k}{R_{OFF}}\int_{t_0}^{t} v_{in}(t)dt' \tag{4.12}$$

Equations (4.11) and (4.12) are used for adjusting the synaptic weights by the use of rectangular voltage or current impulses with suitable durations and polarities [14,46].

For a positive weight tuning, non-negative voltage impulses are needed, and by applying a negative voltage impulse, we can derive negative weight tuning of the corresponding weight [46]. In the preliminary moment, the weight of the memristor-based synaptic scheme has its minimum value w_{min} [46]—Equation (4.13):

$$w_{min} = \frac{M_2}{M_1 + M_2} = \frac{100}{16000 + 100} \approx 0.0062 \qquad (4.13)$$

The maximal possible synaptic weight w_{max} for the memristor circuit is expressed by Equation (4.14) [46]:

$$w_{max} = \frac{M_2}{M_1 + M_2} = \frac{16000}{16000 + 100} \approx 0.9937 \qquad (4.14)$$

The weight range of the suggested circuit could not cover the range needed for expressing the logical functions [46]. By applying a scaling process, the synaptic weight range could be extended and translated to the desirable range for the logical function emulation. The first weight range is $\Delta w_{1max} = 0.9937 - 0.0062 = 0.9875$. The second weight range, applied in the present investigation, has the length: $\Delta w_{2max} = 2 - (-3) = 5$.

The $\Delta w_{2max}/\Delta w_{1max}$ ratio g with a value of 5.0633 is applied for amplification of the output signal of the memristor neural circuit, which is practically the synaptic weight for an input impulse voltage with a level of unity.

After increasing the acquired signal, a summing procedure of the mentioned signal with a constant voltage v_1, described as $v_1 = -3 - 0.0314 = -3.0314$ V, is applied for overlapping of the corresponding weight intervals [46]. The synaptic-weight-adjusting quantity is $\Delta w = 1$ or $\Delta w = -1$. By dividing the $\Delta w_{2max}/\Delta w_{1max}$ ratio of 5.0633, the actual synaptic adjustment for the memristor synaptic circuit is obtained as: $|\Delta w_{real}| = 1/5.0633 = 0.1975$. If a rectangular voltage impulse with an amplitude of 1 V is applied to the input, the impulse duration desirable for adjusting the corresponding synaptic weight is computed [46]—Equation (4.15):

$$\Delta t_i = \frac{\Delta w_{real} R_{OFF}}{k v_m} = \frac{0.197 \times 16000 \times (10^{-8})^2}{10^{-12} \times 100} = 0.0032\,s \qquad (4.15)$$

The same input of the circuit is used for applying the information pulses of the signals x_1 and x_2. To avoid changing the memristors states when the working signals are used, the duration of the logical signals x_1 and x_2 must be many times shorter than the impulse width of the adjusted signals [4,46]. The logical signal amplitude is several times lower than those of the tuning signals. To avoid shifting the boundary between the doped and undoped regions of the memristor element due to the memory effect, the following levels for the input signals x_1 and x_2 are applied: a voltage impulse with a level of 0.2 V for logical unity and a level of -0.2 V for logical zero [46]. The information signals x_1 and x_2 have lengths with a value of 320 μs, that is, about 10 times shorter than the width of the adjusted signals. For applying the information signals and the tuning signals to the memristor-based synaptic input, different time intervals are applied and a multiplexer for sorting out the signals is applied as well [46].

The advantage of the applied memristor-based synapses in the present research is the very small dimensions of the used memristors, their low power consumption and stable non-volatility.

4.2. A Memristor-Based Neural Network and Artificial Neurons

4.2.1. General Information

The neural networks and systems are very important and applicable modules in many electronic schemes and devices, such as telecommunication and electronic circuits, computers and many others [44]. They are frequently used in digital signal processing, image recognition, data clustering, neural computing and many other new areas of the technical industry [37,44]. The traditional neural networks use CMOS technology for the synaptic bonds among the artificial neurons [44]. Software realizations of the synaptic weights are also used [44]. Owing to the high implication of the neural networks, especially their parts—the artificial neurons, the investigation of their new schematic solutions is very supportive for the future generations of the electronic devices, circuits, schemes and networks [37,44]. The memristor element is a promising contender for building the synaptic bonds between the neurons due to its nano-range sizes, low energy consumption, memory effect and the fine compatibility with the present CMOS nanotechnology [14]. To the best of the author's knowledge, there is a definite absence of detailed analyses of neural networks with different memristive synaptic circuits. The motivation for the present research is to fill this absence, offering a precise investigation of a memristor-based linear neural network with one artificial neuron for noise cancellation [37]. The used synapses are memristor-based with only one memristor and three resistors. This synaptic circuit has the possibility for realizing positive, zero and negative synaptic weights in a comparatively broad range. A previously memristor model suggested by the author [33] is applied for the present research.

4.2.2. Mathematical Description of the Applied Memristor Model and the Investigated Artificial Neuron

An adapted bridge synaptic circuit with three nano-scale resistors and a memristor element is suggested and investigated. The bridge synaptic circuits have a specific benefit compared to the single memristor synapses and the memristor-based

135

anti-series synaptic circuit [37]. The bridge schemes [45] are suitable for acquiring negative and zero synaptic weights, and three of the memristors in the bridge circuit are replaced with resistors [37]. For the present research, an improved highly nonlinear memristor model offered by the author in Reference [33] is used. Several main memristor models are also applied for comparison of the obtained results [5,6,14,45]. The ability of the applied memristor model [33] for operation in multipart electronic circuits and schemes is established as well [37].

To augment the nonlinearity of the altered Biolek memristor model for realistic illustration of the nonlinear ionic drift, the author suggests an extra sinusoidal window function component in Reference [33]—System of Equations (4.16):

$$f_{BM}(x) = \left[\frac{1-(x-1)^{2p}}{m+1} + \frac{m\left[\sin^2(\pi x)\right]}{m+1} \right], \qquad v(t) \leq -v_{thr}$$

$$f_{BM}(x) = \left[\frac{1-x^{2p}}{m+1} + \frac{m\left[\sin^2(\pi x)\right]}{m+1} \right], \qquad v(t) > v_{thr}$$

$$\frac{dx}{dt} = k\,i\,f_{BM}(x,i) \tag{4.16}$$

$$\frac{dx}{dt} = 0, \qquad\qquad\qquad -v_{thr} < v \leq v_{thr}$$

$$v = \left[R_{ON}x + R_{OFF}(1-x) \right] i$$

where the variable m is between 0 and 1, and f_{BM} is the improved Biolek window function [33]; the first and the second equations depict the altered window function for different voltage directions; the third equation is the state differential equation of the memristor element; the fourth equation is the state-dependent Ohm's Law. The modified Biolek model applied here is completely defined by System of Equations (4.16). If both the variable m and the sensitivity threshold of the memristor are zero, the altered Biolek model is changed into the standard Biolek model. The flux–charge and the current–voltage characteristics of the memristor element for a sinusoidal voltage signal, according to the Pickett memristor model as a reference [45], are acquired in the PSpice environment [19] and are applied for adjusting the used modified memristor model [33]. The corresponding flux–charge and current–voltage relationships of the applied memristor model [33] are derived after a number of simulations and finetuning of the improved memristor model in accordance to the Pickett model as a reference [7].

The characteristics mentioned above are found by computer analysis, using the numerical solution to System of Equations (4.16) [13,19,37]. After finishing the adjustment processes and obtaining a good matching between the corresponding relationships—the current–voltage and flux–charge relationships, the improved memristor model offered by the author in Reference [33] is applied here for investigation of the considered memristor-based artificial neuron. The basic configuration of the memristor-based artificial neuron under test [37] is shown in Figure 116 for depiction of its structure and the basic signals [37]. This neuron is used for noise cancellation of a low-frequency signal described as $x_1(t) = 0.02 \sin(2\pi \times t + \frac{\pi}{6})$. The desired (target) signal is expressed with $d(t) = 0.03 \sin(2\pi \times t + \frac{5\pi}{18})$. The used activation function in the artificial neuron is linear [39,44]. The output signal derived by the memristor-based artificial neuron $y(t)$ is given in the next equation (4.17) [37,44]:

$$y(t) = x_1 w_1(t) + b w_0(t) \tag{4.17}$$

where x_1 is the input signal and b is the shift coefficient [37]. The quantities $w_1(t)$ and $w_2(t)$ are the synaptic weights of the corresponding synapses of the artificial neuron. The error signal $e(t)$ acquired by the memristor neuron is expressed by Equation (4.18) [44]:

$$e(t) = -y(t) + d(t) \tag{4.18}$$

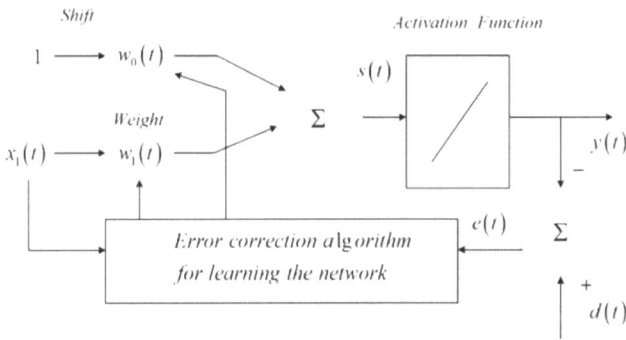

Figure 116. A schematic of a linear neuron for noise cancellation, presented for clarification its basic configuration and the applied signals.

The error correction algorithm for learning the neuron is based on tuning of the corresponding synaptic weight coefficients—Equation (4.19) [44]:

$$w(t + \Delta t) = w(t) + \eta_1 e(t) x_1(t + \Delta t) \tag{4.19}$$

where η_1 with a value of 0.05 is the used learning rate [37,44]. The synapses in the neural network are memristor-based and their circuit is shown in Figure 117.

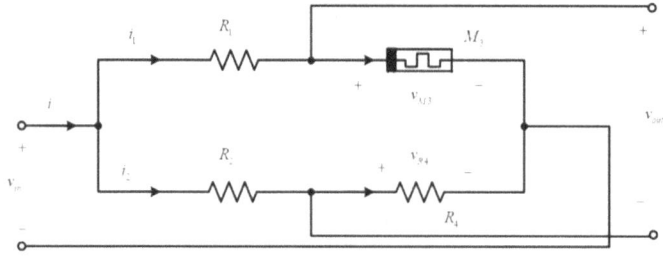

Figure 117. A memristor-based bridge synaptic device with three nano-scale resistors and a memristor component.

The synaptic scheme illustrated in Figure 117 contains three nano-scale resistors with dimensions similar to the memristor size and a memristor element. The bridge circuit makes it possible to acquire negative synaptic weights [37,45]. After investigation of the synaptic circuit, the weight of the circuit w is derived from the following Equation (4.20) [37]:

$$w = \frac{v_{out}}{v_{in}} = -\frac{R_4}{R_2 + R_4} + \frac{M_3}{R_1 + M_3} \tag{4.20}$$

where the memristance of the element M_3 is given in Equation (4.21) [45]:

$$M_3 = R_{ON}x + R_{OFF}(1 - x) \tag{4.21}$$

The sign and value of the synaptic weight w are functions of the resistances R_1, R_2, and R_4 and the memristance M_3. The synaptic weight could be changed by the use of voltage impulses with an amplitude of 2 V and different polarities and durations, dependent on the desired synaptic weight [37,45].

The main advantages of the proposed synaptic circuit by the author (Figure 117), with respect to the synapses shown in Figure 113, is the possibility for obtaining zero and negative synaptic weights. It is established that the range of the synaptic weight for the present memristor circuit is from -0.33 to 0.47 and it is able to cover the needed interval for the present investigation. The disadvantage of the described circuit (Figure 117), with respect to the bridge circuit in Reference [45] which has synaptic change in the interval $(-1, 1)$, is the narrower interval for changing the synaptic weight. An advantage of the presented circuit with respect to the bridge circuit in Reference [45] is the reduced number of the used memristors. The electric current flowing through the memristor M_3 in the tuning time intervals is given in Equation (4.22) [37]:

$$i_{M3} = \frac{v_{in}}{M_3 + R_1} = \frac{v_{in}}{(R_{ON} - R_{OFF})x + (R_1 + R_{OFF})} \tag{4.22}$$

Using Equations (4.21) and (4.22), the state differential Equation (4.23) for the memristor element in the synaptic circuit is derived:

$$\frac{(-R_{OFF} + R_{ON})x + (R_{OFF} + R_1)}{f_{MB}(x)} dx = k\, v_{in}\, dt \qquad (4.23)$$

Applying System of Equations (4.16), and Equations (4.20) and (4.23), a pseudo-code is obtained for their numerical solution according to the relationship between the synaptic weight w and the duration of the adjusted pulses. The suggested memristor neuron is first analyzed in MATLAB [13,37] using an algorithm for investigation of the idealized circuit. Then, it is established that the weight changes in a range from -0.3 to 0.3.

The memristor-based artificial neuron is analyzed, using the suggested memristor model [33] and several main existing models as well [5,6,14]. A reasonably good similarity between the obtained results is established. The durations of the adjusted impulses are different for the different iterations and are numerically calculated in accordance to the desired synaptic weights. The input signal applied for noise suppression has a magnitude lower than the sensitivity threshold of the memristor element and it does not change the synaptic weights [14,33].

The basic parameters, characteristics and constants, connected to the present investigation, are presented in Table 4.

Table 4. Parameters and constants of the used elements and the memristor model.

Parameter	Explanation	Numerical Value	SI Unit
R_{ON}	ON-resistance of M_3	100	Ω
R_{OFF}	OFF-resistance of M_3	16,000	Ω
R_1	Resistance R_1	500	Ω
R_2, R_4	Resistances of R_2 and R_4	100	Ω
v_{in}	Adjusting signal level	2	V
μ	Oxygen vacancies mobility	1×10^{-14}	$m^2/(V{\cdot}s)$
D	Memristor element length	10×10^{-9}	m
η	Polarity (biasing) coefficient	1	-
p	Positive integer exponent in the altered window function	10	-
m	Weight variable in the altered window function	0.2	-
η_1	Learning rate of the memristor-based artificial neuron	0.05	-

4.2.3. Results and Discussion

The time graphs of the input signal, the desired (target) and the output signals after training the artificial memristor neuron are presented in Figure 118 for additional explanations. It is clear that a fine matching between the desired (target) and the output signals is established. The neuron successfully suppresses the noise signals.

The time graphs of the memristance and the state variable x_3 of the memristor M_3 in the working process are shown in Figure 119(a,b) for describing the state alteration and the adjustment processes of the synapses. The normal value of the resistance of the memristor element is 500 Ω. The alteration of the memristance in this situation is about several ohms. The corresponding value of the state variable of the memristor element is about 0.97 and its change is in a very narrow interval [37].

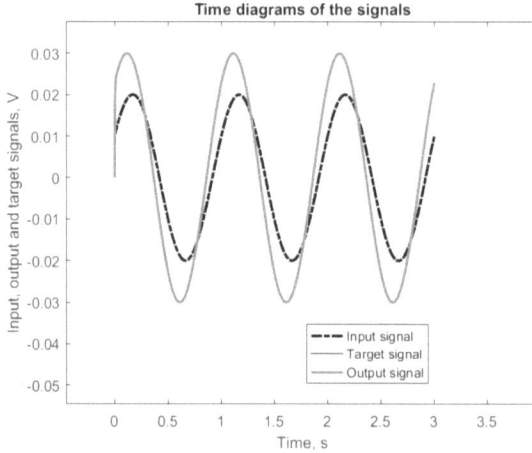

Figure 118. Time graphs of the input signal, the output signal and desired (target) signal, acquired by the memristor-based artificial neuron for noise suppression. A good overlapping between the desired (target) and the output signals is realized. The artificial neuron effectively suppresses the noise signals. The input signal is described as: $x_1(t) = 0.02 \sin\left(2\pi \times t + \frac{\pi}{6}\right)$. The output and desired (target) signals are the same, expressed as: $d(t) = 0.03 \sin\left(2\pi \times t + \frac{5\pi}{18}\right)$.

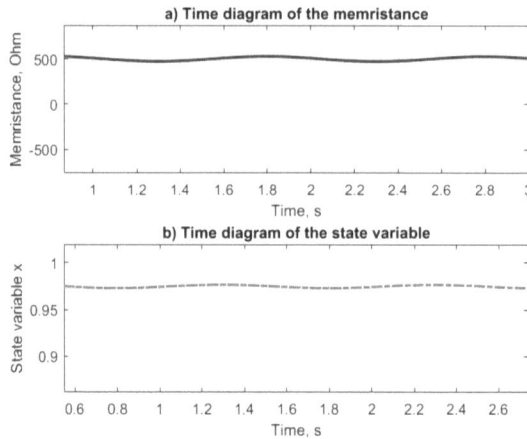

Figure 119. (a) Time graph of the memristance of the memristor; (b) Time diagram of the state variable of the memristor element in the memristor-based artificial neuron.

140

After discussing the results, it could be established that the suggested altered memristor-based synaptic scheme could be successfully used in artificial neurons and neural networks for noise suppression. The used memristor model previously offered by the author in Reference [33] is very appropriate for analyzing memristor-based neural networks, because it has an activation threshold for low-voltage signals and a strong nonlinearity for realistic illustration of the nonlinear ionic drift in the memristor elements. The described neuron could be used in multilayer neural networks for different purposes.

4.3. Learning a Neuron with Resistor–Memristor Synapses

4.3.1. General Information

Due to the high importance of the artificial neurons, synthesis and analysis of their new well-organized circuit solutions are valuable for the future generation of electronic systems and devices [44]. The memristor element is a promising candidate for building the synaptic connections between the neurons due to its nano-scale dimensions, low power consumption, memory effect and the compatibility with the current CMOS nanotechnology [14]. To the best of the author's knowledge, there is a certain absence of detailed analysis of artificial neurons with different memristor-based synaptic schemes. The motivation for the present analysis is to fill this lack, offering a detailed investigation of a memristor-based linear neuron for noise suppression with memristor bridge synapses with two memristors and two nanoscale resistors [39]. In a previous proposal [37], a neuron with synaptic circuits containing only one memristor element is investigated. For the present research, a modified bridge synaptic circuit with two nano-scale resistors and two memristors is offered and analyzed [39]. A different modified highly nonlinear memristor model offered by the author in Reference [35] is used as well. Several basic memristor models are applied for a comparison of the basic significant characteristics [5,6,14]. The ability of the used memristor model [35] for operation in complex neural electronic circuits is established.

4.3.2. Mathematical Description of the Used Memristor Model and the Memristor-Based Artificial Neuron

The main state differential equation of the memristor element [14] is given in the next Equation (4.24):

$$\frac{dx}{dt} = \eta \, k \, i \, (x, v) \, f(x) = \eta \, \frac{\mu \, R_{ON}}{D^2} \, i \, f(x) \tag{4.24}$$

where η is a polarity coefficient; for a forward-biasing, it has a value of 1, and for reverse-biased memristor element, it is equal to -1; k is a constant, dependent on memristor physical characteristics, μ is the ionic mobility of the oxygen vacancies in

TiO$_2$ and equal to 1×10^{-14} m^2/(V·s), and $f(x)$ is a window function [5,6,14] used for representation of the nonlinear dopant drift and the boundary effects. The Joglekar window function is used in Reference [5] and it is expressed by Equation (4.25):

$$f(x) = f_J(x) = 1 - (2x - 1)^{2p} \tag{4.25}$$

where p is a positive integer exponent and x is the memristor state variable. Another very significant window function with extensive application [6], first presented by Biolek in Reference [6] and also known as a Biolek window function [6,7], is given in Equation (4.26):

$$f(x) = f_B(i, x) = 1 - [x - stp(-i)]^{2p} \tag{4.26}$$

where $stp(i)$ is the relay function dependent on the current direction and can be written in the next Equation (4.27):

$$stp(i) = \begin{cases} 1, & if\ i \geq 0 \quad (v \geq 0) \\ 0, & if\ i < 0 \quad (v < 0) \end{cases} \tag{4.27}$$

After substituting $stp(i)$ into Equation (4.26), a different expression of the Biolek window function is acquired [6,7]—Equation (4.28):

$$f_B(x) = 1 - \left[(x-1)^{2p}\right], \quad i(t) \leq 0$$

$$f_B(x) = 1 - x^{(2p)}, \qquad i(t) > 0 \tag{4.28}$$

The memristor model applied for the analysis in the present research is offered by the author and described in details in Reference [35]. The modified window function $f_M(x)$ proposed in Reference [35] is based on both Joglekar [5] and Biolek [6] window functions. It is a simple combination of these two window functions. Hence, the used final window function is presented with Equation (4.29) [35]:

$$f_M(x, v) = \frac{f_J(x) + f_B(x, i)}{2} \tag{4.29}$$

After substitution of Equations (4.25)–(4.28) in (4.29), a more suitable expression of the suggested altered window function is obtained—Equation (4.30) [35]:

$$f_M(x) = 1 - \frac{(x-1)^{2p} + (2x-1)^{2p}}{2}, \quad i(t) \leq 0$$

$$f_M(x) = 1 - \frac{x^{2p} + (2x-1)^{2p}}{2}, \qquad i(t) > 0 \tag{4.30}$$

The depiction of the nonlinearity of the memristor ionic drift could be expressed by decreasing the integer exponent [5,6,14] in the suggested improved window function. There are many possible formulae for illustrating such a function. The author attempted to apply a simplified variant for this relationship to optimize the memristor model performance. The applied relationship between the integer exponent p and the applied voltage v is Equation (4.31) [35]:

$$p = round\left(\frac{a}{|v| + c}\right) \tag{4.31}$$

where the special function "*round*" is applied for deriving the integer result; the quantities a and c could be roughly acquired by comparing the results with these obtained by the Pickett model and adjusting the altered memristor model; the constant c is used for avoiding the division by zero [35].

The suggested improved window function $f_M(x,v)$, shown in System of Equations (4.30), is substituted into the state-dependent current–voltage characteristic, and the suggested memristor model could be depicted by System of Equations (4.32) [35]:

$$\frac{dx}{dt} = \eta \, k \, i \left\{ 1 - \frac{1}{2}\left[\begin{array}{l} (x-1)^{2\cdot round\left(\frac{a}{|v|+c}\right)} + \\ +(2x-1)^{2\cdot round\left(\frac{a}{|v|+c}\right)} \end{array} \right] \right\}, \quad v(t) \leq 0$$

$$\frac{dx}{dt} = \eta \, k \, i \left\{ 1 - \frac{1}{2}\left[\begin{array}{l} x^{2\cdot round\left(\frac{a}{|v|+c}\right)} + \\ +(2x-1)^{2\cdot round\left(\frac{a}{|v|+c}\right)} \end{array} \right] \right\}, \quad v(t) > 0 \tag{4.32}$$

$$v = R \, i = [R_{ON} x + R_{OFF}(1 - x)] \, i$$

It is established that for $a = 30$ and $c = 2$, the corresponding current–voltage and the flux–charge relationships are almost identical to those acquired by the use of the reference Pickett model. The current–voltage and the flux–charge relationships of the memristor are multi-valued functions. During the analyses of Pickett model [7], when voltages higher than 0.75 V are applied, many computational problems occur. The main advantage of the suggested modified model [35] compared to the standard Pickett memristor model is the absence of convergence issues.

The configuration of the memristor-based artificial neuron under investigation [39] is illustrated in Figure 120 for explanation of the basics of learning and adjusting the artificial neuron. It is used for noise suppression of a low-frequency signal described as $x_1(t) = 0.03 \sin(2\pi \times 2t + \frac{\pi}{9})$. The desired (target) signal is denoted by $d(t) = 0.07 \sin(2\pi \times 2t + \frac{5\pi}{18})$ [39]. The error correction algorithm for learning and tuning the neuron is based on adjustment of the weight coefficients of the synaptic bonds [44]—Equation (4.33):

$$w(t) = w(t - \Delta t) + \eta_1 x_1(t) e(t) \tag{4.33}$$

where η with a value of 0.03 is the used learning rate of the neuron [39,44]. The synapses are memristor-based and their schematic is shown in Figure 121 [39] for further explanations of the basics of the neuron synapse operation processes and the input and output signals. The memristor elements are connected in an anti-parallel biasing manner [39].

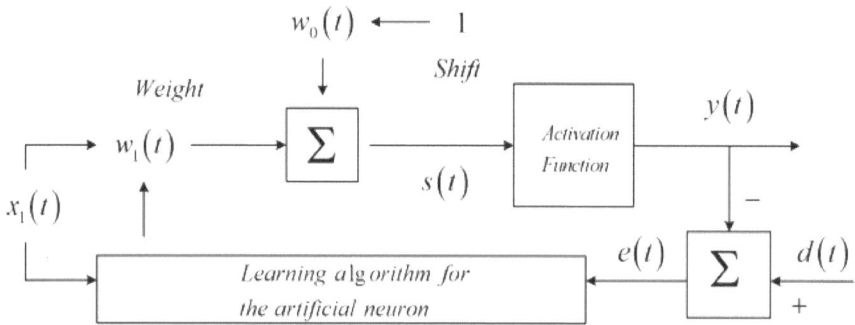

Figure 120. A schematic for illustration of the neuron learning processes. The input signal is $x_1(t) = 0.03 \sin(2\pi \times 2t + \frac{\pi}{9})$, and the desired (target) signal is $d(t) = 0.07 \sin\left(2\pi \times 2t + \frac{5\pi}{18}\right)$.

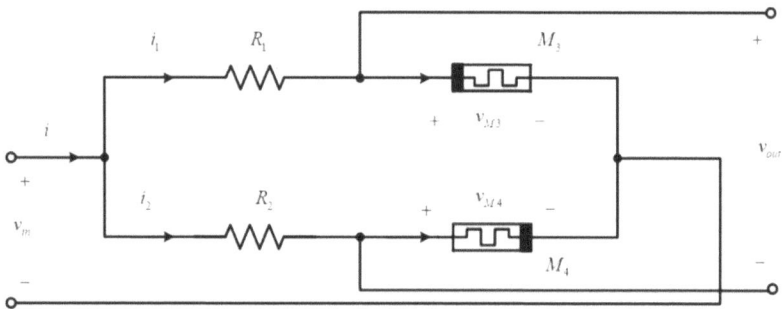

Figure 121. A memristor–resistor synaptic scheme with two nano-scale resistors and two memristors, realized in a bridge circuit.

The synaptic scheme shown in Figure 121 contains two nano-scale resistors with sizes similar to the memristor element size, and two titanium dioxide memristor components [39]. The bridge schematic makes it possible to acquire negative synaptic weights because the output voltage of the circuit is the difference between the voltages

across the memristors [33,39]. After investigation of the synaptic scheme, the output
signal and the weight of the synapse w are obtained [39]—Equation (4.34):

$$v_{out} = \left(-\frac{M_4}{R_2+M_4} + \frac{M_3}{R_1+M_3}\right)v_{in}$$

$$w = \frac{v_{out}}{v_{in}} = -\frac{M_4}{R_2+M_4} + \frac{M_3}{R_1+M_3}$$

(4.34)

where the memristances of the elements M_3 and M_4 and the weight w are presented
in the System of Equations (4.35):

$$M_3 = R_{ON}x_3 + R_{OFF}(1 - x_3) \qquad w = \frac{R(M_3 - M_4)}{(R + M_3)(R + M_4)}$$
$$M_4 = R_{ON}x_4 + R_{OFF}(1 - x_4)$$

(4.35)

The sign and value of the synaptic weight w depend on the resistances R_1, R_2,
and the memristances of the elements M_3 and M_4 [39]. The synaptic weight could be
altered by the application of voltage pulses with amplitude of 2.4 V and different signs
and impulse durations, dependent on the corresponding target synaptic weight [39].
It is established that the range of the synaptic weight for the present memristor-based
scheme is from -0.49 to 0.49. The memristor synapse shown in Figure 113(b) has
also two memristors but it cannot represent zero and negative weights, owing to
the polarities of the input and the output voltages. An advantage of the memristor
synapse shown in Figure 121, compared to that illustrated in Figure 117 [37], is the
broader interval for change of the synaptic weights. Another advantage with respect
to the bridge synapse in Reference [45] is the reduced number of the used memristors
per synapse. A disadvantage of the discussed synapse (Figure 121) with respect to the
synapse proposed in Reference [45] is the narrower range for change of the synaptic
weights. This established broad range of altering the synaptic weight is a benefit of
the applied synaptic scheme with two memristors compared to the corresponding
circuit suggested in Reference [37]. The currents flowing through the memristors M_3
and M_4 in the tuning time intervals are given by System of Equations (4.36) [39]:

$$i_{M3} = \frac{v_{in}}{M_3+R_1} = \frac{v_{in}}{(R_{ON}-R_{OFF})x_3+(R_{OFF}+R_1)}$$

$$i_{M4} = \frac{v_{in}}{M_4+R_2} = \frac{v_{in}}{(R_{ON}-R_{OFF})x_4+(R_{OFF}+R_2)}$$

(4.36)

Using System of Equations (4.36), the state differential equations for the memristors are acquired—System of Equations (4.37):

$$\frac{(R_{ON}-R_{OFF})x_3+(R+R_{OFF})}{f_M(x_3)}d\,x_3 = k\,v_{in}\,dt$$

$$\frac{(R_{ON}-R_{OFF})x_4+(R+R_{OFF})}{f_M(x_4)}d\,x_4 = -\,k\,v_{in}\,dt$$

(4.37)

The basic parameters and constants, associated with the present analysis, are shown in Table 5.

Table 5. Parameters of the elements and the used memristor model.

Quantity	Determination	Numerical value	SI unit
R_{ON}	ON-resistance of M_3	100	Ω
R_{OFF}	OFF-resistance of M_3	16,000	Ω
R_1, R_2	Resistance of R_1 and R_2	100	Ω
v_{in}	Adjusting signal level	2.4	V
μ	Oxygen vacancies mobility	1×10^{-14}	$m^2/(V \cdot s)$
D	Memristor size	10×10^{-9}	m
η_3, η_4	Polarity coefficients of the elements M_3 and M_4	$1; -1$	-

4.3.3. Results and Discussion

It is established that the synaptic weights in the noise cancellation process changes in the range from -0.06 to 0.06. The corresponding range acquired by the present memristor synapse completely covers the needed interval and no extra scaling components are needed. The memristor-based artificial neuron is investigated using the applied model [35] and several existing basic models as well [5,6,14]. A reasonably good similarity between the results obtained by these models is established. The durations of the adjusted voltage pulses are different for the different iterations and are calculated in dependence on the weights [39]. A good overlapping between the desired (target) and the output signals is established (Figure 122).

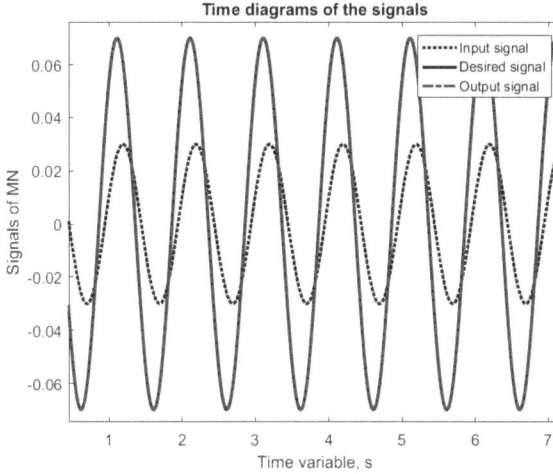

Figure 122. Time graphs of the input signal, the output signal and the desired signal, acquired by the memristor neuron for noise suppression. A good coincidence between the desired (target) and the output signals is established; the neuron effectively suppresses the noise signals. The input signal is $x_1(t) = 0.03\sin(2\pi \times t + \frac{\pi}{6})$, and the desired (target) signal is $d(t) = 0.07\sin\left(2\pi \times t + \frac{5\pi}{18}\right)$.

The input signal used for noise filtering has a level lower than the activation threshold of the memristor element and does not affect the synaptic weights [39]. It could be established that the suggested altered memristor synaptic scheme could be successfully applied in artificial neurons [35,39]. The analyzed neurons could be used in multilayer neural networks.

4.4. A Passive Memristor Matrix

4.4.1. General Information

The memristor memory technology is promising, which could potentially replace the conventional CMOS and flash-based non-volatile memory-integrated circuits and devices [25,38,47,48].

The memristor element is used in the memory schemes as a storing element [47,48]. To the best of the author's knowledge, there is a certain lack of detailed and complete results derived by memristor memory analysis with the basic memristor models. The motivation for the present research is to fill this absence, suggesting detailed investigation of a passive memristor memory fragment, using an altered highly nonlinear memristor model suggested by the author in Reference [33], and making a comparison of the current–voltage characteristic derived by the proposed model with experimentally recorded current–voltage characteristics of memristor elements in memory crossbar circuits.

A comparison of the derived results with these acquired by the application of Generalized Boundary Condition Memristor (GBCM) model [14] is completed as well. The ability of the suggested memristor model [33] for realistic illustration of the behavior of complex memristor-based electronic schemes is established.

4.4.2. Mathematical Description of the Applied Memristor Model and the Basic Processes in the Passive Memristor Memory Matrix

The adapted author's memristor model applied for the computer simulations in the present investigation is illustrated in details in Reference [33].

A plain modification of the standard Biolek window function is realized in Reference [33]. For growing the nonlinearity of the proposed altered Biolek model, the author suggests an additional sinusoidal window function term—Equation (4.38) [33]:

$$f_{BM}(x) = \left[\frac{1-(x-1)^{2p}+m[\sin^2(\pi x)]}{m+1} \right], \quad v(t) \leq 0$$

$$f_{BM}(x) = \left[\frac{1-x^{2p}}{1+m} + \frac{m[\sin^2(\pi x)]}{1+m} \right], \quad v(t) > 0$$

(4.38)

where the coefficient m is between 0 and 1, and f_{BM} is the suggested modified Biolek window function [33]. The improved Biolek memristor model applied here is fully described with the next System of Equations (4.39) [33]. The optimum value for the coefficient m has been acquired with a value of 0.2 [33].

$$\frac{dx}{dt} = \eta\, k\, i \left[\frac{1-(x-1)^{2p}+m\left(\sin^2(\pi x)\right)}{1+m} \right], \qquad v(t) \leq 0, \quad i(t) \leq 0$$

$$\frac{dx}{dt} = \eta\, k\, i \left[\frac{1-x^{2p}+m\left(\sin^2(\pi x)\right)}{1+m} \right], \qquad v(t) > 0, \quad i(t) > 0$$

(4.39)

$$v = R\, i = [(R_{ON} - R_{OFF})\, x + R_{OFF}]\, i = [\Delta R\, x + R_{OFF}]\, i$$

A fragment of memristive memory matrix with 16 memory cells according to Reference [48] is illustrated in Figure 123 for the following explanation of the signals used for writing and reading procedures. The electrical circuit of the memory fragment is shown in Figure 124 for describing the corresponding working procedures. It contains four rows (bit lines) and four columns (word lines) [38]. Selecting the particular memory component makes possible loading a bit of information—logical unity or logical zero.

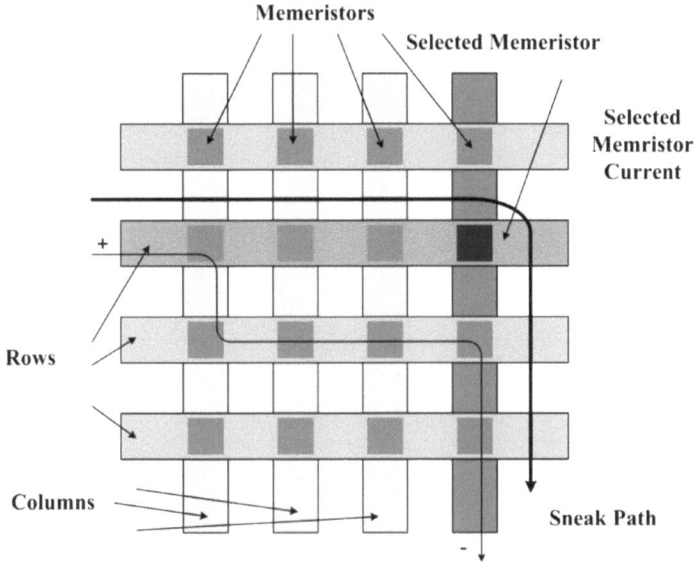

Figure 123. A fragment of a memristor matrix representing the rows, the columns, the memristor memory cells, the selected memristor current path and a parasitic sneak path.

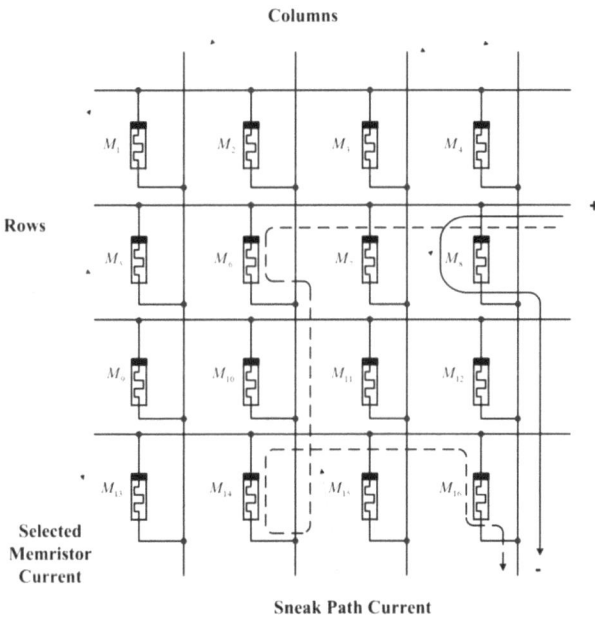

Figure 124. A simplified substituting electric circuit of a fragment of a passive memristor-based memory crossbar.

The resistance of the rims between the neighboring memristor elements is about 1.25 Ω [38] and it could be neglected according to the minimum memristance of the memory cells. For writing logical unity, a positive pulse with a level of 2 V is applied to the corresponding memristor. For writing logical zero, a negative voltage signal with the same amplitude is applied [38]. For the reading processes, a positive voltage signal with a value lower than the memristor sensitivity threshold is used [14,33].

Memristor models with activation thresholds, such as the GBCM model and the applied nonlinear memristor model by the author [33], are appropriate for investigation of the reading processes in passive memory matrices. In the first 100 ms, logical unity is stored in the memristor cell M_1. In the next 100 ms, a reading signal with a level of 200 mV is applied to the memristor element M_1. This level is lower than the sensitivity threshold of the memristor element, and during the reading process, the information accumulated in the memristor will not be affected. The voltage across a series-connected sense resistor with a resistance of 1788 Ω [38] is proportional to the logical level stored in the corresponding memristor cell [38,47]. In the next 100 ms, a negative impulse with an amplitude of 2 V is used and logical zero is written in the memristor memory cell [25,38].

The next procedure is reading the information from the memory cell for 100 ms. A parasitic sneak path between the electrodes is illustrated in Figures 123 and 124. It has been specified by a number of analyses and a comparison with a single-cell memory that, in this case, the sneak paths do not strongly affect the normal operation of the memory device [38]. This fact could be explained with the occurrence of a reverse-biased memristor element with a very high resistance in the sneak path and the corresponding rectifying effect related to the operation of the memristors in a regime close to a hard-switching mode [14,38].

4.4.3. Results and Discussion

For selecting the memristor element M_1, a positive potential with a value of 2 V is applied to the first row electrode of the crossbar. The first column electrode is grounded [38,47]. The time graphs of the memristor voltage, the resistance of the memristor M_1 and the output voltage taken after a reading procedure are presented in Figure 125 for further explanation of the device operation. They characterize the range of the resistance of the memristor element and the result obtained by applying the reading signal [38]. The resistance of the corresponding memory element changes in a very wide range from 200 Ω (for logical unity) to about 12 kΩ (for logical zero).

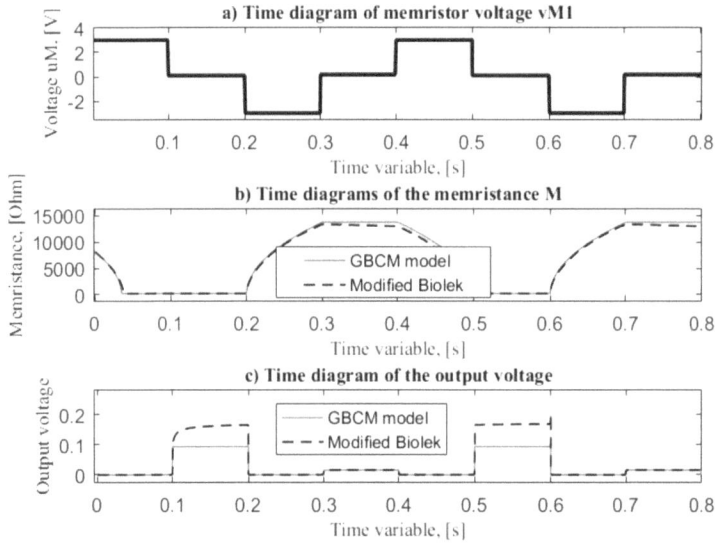

Figure 125. (**a**) Time graph of the memristor voltage; (**b**) Time diagram of the memristance, according to the GBCM and the modified Biolek model; (**c**) Time digram of the output voltage, according to the GBCM and the modified Biolek model and derived by applying the reading procedure.

The comparatively wide range of variation of the memristance is valuable for the precise recognition of the logical levels, because the output voltage depends on the memristance derived immediately after the writing process. The GBCM model and the altered Biolek model are used for the present investigations of the derived results comparison. The corresponding time diagrams of the memristance almost match each other.

An advantage of the suggested memristor model is that the output voltage acquired by the use of the improved Biolek model for the reading process is slightly higher than the corresponding voltage signal acquired by the GBCM model [14,38]. It is established by additional investigations that low variations of the main memristor quantities do not strongly affect the normal operation of the memory crossbar.

The current–voltage and memristance–flux characteristics of the memristor M_1 derived in the operation process of the memory scheme are presented in Figure 126 and they illustrate the memristor performance in the corresponding operating regime. The current–voltage relations, which are basic characteristics of the memristor element for a given signal, practically matchingeach other.

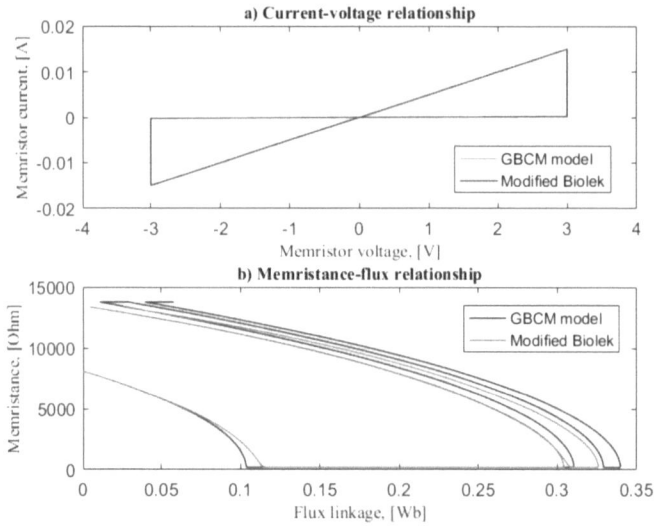

Figure 126. (a) Current–voltage and (b) memristance–flux characteristics of the memristor M_1 according to the applied model and GBCM model, derived during the memory operation.

The resistance of the memristor element almost attains its limiting values. The memristor element operates in a state close to a hard-switching regime [38]. After comparing the current–voltage characteristics of the memristor derived by the use of the applied model [33] with an experimental current–voltage relationship [47,48], a good similarity between these two current–voltage functions is established. The results acquired by the application of the suggested memristor model are identical to the experimentally recorded characteristics [25,49] and the results derived by the use of GBCM model [14,38]. The respective memristance–flux characteristics derived by the use of GBCM model and the altered Biolek model are similar and they almost overlap with each other [38].

The output voltage of the memory circuit is derived during the reading procedure by the application of a resistor and an amplifier [38]. The output voltage has a comparatively high level, which is a good precondition for a precise differentiation of the logical levels. Comparing the corresponding current–voltage relationship of the memristor element with experimental data [47,48] derived under similar conditions and with the results acquired by the GBCM memristor model, their good similarity is obtained. It could be concluded that the suggested nonlinear memristor model [33] could be applied for analysis of many complex memristor-based electronic schemes and devices. An advantage of the memristor model used here, compared to the GBCM model, is the reasonably high nonlinearity extent of the ionic dopant drift [38].

The established good convergence of the computational procedures using the corresponding algorithm for the computer simulations [33] is also an advantage of the applied model [38]. The successful operation of a memristor memory matrix with the suggested memristor model for the general impulse mode confirms that this model could be used for analysis of many memristor-based electronic circuits and devices [38]. After analyzing the derived results, it could be concluded that the applied memristor models have almost similar behavior in the operation procedures: writing and reading logical information [38].

The respective current–voltage relationship and memristance–flux characteristics derived by the memristor model offered by the author in Reference [33] and the GBCM model are identical and practically they match one to another. The model used here has the ability to represent the behavior of a memristor in memory circuits for the general impulse regime. With a comparison of the current–voltage characteristics derived by the proposed model with experimentally recorded current–voltage relationship, it could be concluded that the model used there could realistically illustrate the behavior of the memristor elements in multipart schemes. The memristor model proposed in Reference [33] could be effectively applied for representation of writing procedures in a memristor memory matrix when the memristors operate in a state close to a hard-switching regime and for the reading processes with a voltage lower than the activation thresholds of the memristor elements.

4.5. A Hybrid Resistance-Switching Memory Device with Memristors

4.5.1. General Information

The analysis of new memory schemes is important for upcoming generation electronic circuits and devices [40,47–49]. The goal of this investigation is to present a detailed analysis of a resistance-switching memory fragment with memristors and separate MOS transistors [40]. The hybrid MOS–memristor technology is a very significant new part of electronics [40,48]. This technology could potentially alter the traditional memory chips [40,49]. The memristor element is applied in the hybrid memories as a storing component [40]. The MOS transistors are applied for removing the sneak path problems existing in the passive memristor memory matrices [40,47,48]. To the best of the author's knowledge, there is a certain lack of comprehensive results acquired by memristor memory physical measurements and analysis with the main memristor models. The motivation for the present investigation is to fill this deficiency, offering extra detailed research of a section of a hybrid memory scheme [40]. For this analysis, a highly nonlinear model [40], and a modified nonlinear window function, suggested by the author in Reference [35], are used. A comparison of these results with experimental data is done as well. The capability of the used memristor model [21,22,25,40] with the improved window function [35] for realistic depiction of the behavior of multipart memristor schemes for soft- and hard-switching modes is established.

4.5.2. Mathematical Description of the Applied Memristor Model and the Processes in the Memristor-Based Memory Crossbar with MOS Transistors

The nonlinear current–voltage relationship of the memristor is advantageous for logic circuits, and therefore more suitable memristive device models without convergence issues have been proposed [25]. In References [20–22], a physical memristor model is illustrated, based on the experimental data in References [20,21]. The approximated relationship between the memristor current i and the applied voltage signal v is presented in Equation (4.40) [20,21]:

$$i = \beta x^n \sinh(\alpha v) + \chi[-1 + \exp(\gamma v)] \tag{4.40}$$

where α, β, γ and χ are tuning parameters, and n is a parameter determining the influence of the state variable x on the memristor current i. In this memristor model [20,21], the state variable x is a normalized parameter in the interval [0, 1]. This model illustrates an asymmetric switching behavior. When the memristor element is in the ON-state, the state variable is close to unity and the electric current is dominated by the first term in Equation (4.36), which depicts a tunneling effect [20,26]. When the memristor element is in the OFF-state, the state variable is close to zero and the current is mainly expressed by the second term in Equation (4.36), which is similar to semiconductor diode current–voltage characteristic [35]. The used memristor model [20,21] applies a nonlinear dependence on the memristor voltage v in the state differential Equation (4.41) [20,21]:

$$\frac{dx}{dt} = f(x) \, a \, v^m \tag{4.41}$$

where a is a constant, m is an odd integer exponent, and $f(x)$ is a window function used for rough illustration of the nonlinear dopant drift and the boundary effects [14,35,40]. The used window function introduces nonlinearity according to the state of the memristor element [14,41]. Equations (4.40) and (4.41) classify the corresponding physics-based memristor model [20,21]. The ionic transport is associated to the ionic drift in the corresponding memristor material [20,25]. The ions jump between two neighboring states via a migration wall [20,25]. This potential barrier could be decreased by the applied external electric field. The ions can derive more thermal energy by heating and can easily overcome the tunnel wall. The nonlinearity of the ionic drift starts from local Joule heating or higher electric fields [20,25]. The applied window function [35,40] gives an approximate relationship between the state variable and the flowing electric current [5,6,14]. The improved window function $f_M(x)$ proposed in Reference [35] is based on both Joglekar [5] and Biolek [6] window functions. It is a linear combination of these window functions with the maximum value of unity and the corresponding minimum value of zero, can be written as the next Equation (4.42) [35,40]:

$$f_M(x,v) = \frac{f_J(x) + f_B(x,i)}{2} \tag{4.42}$$

A more suitable form of the altered window function is presented in Equation (4.34) [35,40]:

$$f_M(x) = -\frac{(x-1)^{2p}+(2x-1)^{2p}}{2} + 1, \quad i(t) \leq 0$$

$$\tag{4.43}$$

$$f_M(x) = -\frac{x^{2p}+(2x-1)^{2p}}{2} + 1, \quad i(t) > 0$$

If the applied voltage increases, the ionic drift nonlinearity increases as well [5,6,40]. The representation of the alteration of the nonlinearity of the dopant drift could be illustrated with the decrease of the integer exponent in the improved window function [35]. There are many potential equations for representing this relationship. The author tried to suggest a simple variant for this function to optimize the performance and reduce the computational time. The applied relationship between the integer exponent p and the applied memristor voltage v [35] is expressed by Equation (4.44):

$$p = round\left[\frac{a}{c + |v|}\right] \tag{4.44}$$

where the specialized function "round" is applied for deriving an integer result; The quantities a and c could be acquired by comparing the results of the suggested altered model with those derived by the Pickett model and then tuning the altered model. The constant c is applied for avoiding division by zero if the applied voltage has a value of zero. The altered window function $f_M(x,v)$ [35] is substituted into the equation describing the current–voltage memristor relationship. The used memristor model [20,22,35,40] could be illustrated with the following System of Equations (4.45) [40]:

$$\frac{dx}{dt} = a\left\{1 - \frac{1}{2}\left[\begin{array}{c}(x-1)^{2\cdot round(\frac{a}{|v|+c})}+\\+(2x-1)^{2\cdot round(\frac{a}{|v|+c})}\end{array}\right]\right\}v^m, \quad v(t) \leq 0$$

$$\tag{4.45}$$

$$\frac{dx}{dt} = a\left\{1 - \frac{1}{2}\left[\begin{array}{c}x^{2\cdot round(\frac{a}{|v|+c})}+\\+(2x-1)^{2\cdot round(\frac{a}{|v|+c})}\end{array}\right]\right\}v^m, \quad v(t) > 0$$

$$i = \chi[\exp(\gamma v) - 1] + \beta x^n \sinh(\alpha v)$$

155

where the third equation depicts the state-dependent Ohm's Law for the memristor element [11]. The analysis of the memristor element is made using numerical solution to System of Equations (4.45) in accordance to the finite difference method. After many computer simulations of the applied memristor model for different values of the quantities a and c, it is established that for $a = 30$ and $c = 2$, the current–voltage characteristic is almost identical to this acquired by the Pickett model [7]. The current–voltage characteristic of the memristor is a multi-valued function. For the simulations using the Pickett model as a reference [7,20] with voltages higher than 0.75 V, many convergence issues occur [41]. The basic benefit of the suggested model [20,22,40] compared to the Pickett model is the absence of computational problems. After finishing the tuning procedures and deriving a reasonably good similarity of the current–voltage relationships between the suggested model and the Pickett model with respect to their current and voltage ranges and current–voltage relationship forms, the suggested memristor model is applied for analysis of the resistance-switching memory [40].

A fragment of a resistance-switching memory scheme with four memristor cells and several separate MOS transistors according to [47,48] is shown in Figure 127 for the following description n of the signals for writing and reading processes. Selecting the corresponding memory elements makes it possible to store a bit of information—logical unity or logical zero in the memory element [47,48]. The MOS transistors are applied for eliminating the parasitic sneak paths between the bit rims and the corresponding word lines [47]. The "write enable" and "read enable" signals are applied to the gate electrodes of the respective MOS transistors and the target memory element will be selected [40,47,48]. For writing logical unity, a positive voltage impulse is applied to the corresponding memristor. For writing logical zero, a negative signal is used [40,48].

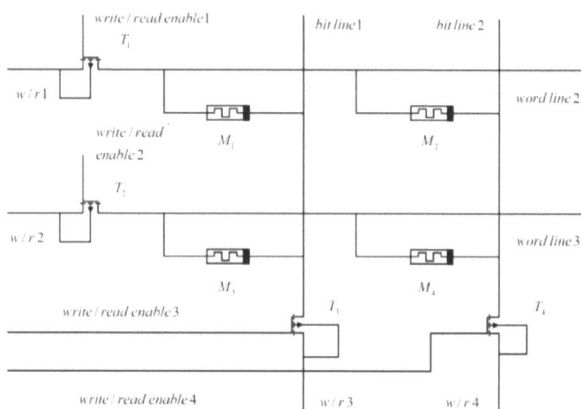

Figure 127. A fragment of a hybrid memristive resistance-switching memory circuit.

For analysis of the hybrid memory circuit fragment, a nonlinear physics-based memristor model [20,22] with an improved window function by the author [35] is introduced. The potentials of the source electrodes of the MOS transistors T_1 and T_3, needed for the processes of writing and reading logical information, are impulses with different levels and directions. In the first 100 ms, logical unity is stored in the memristor element M_1. In the next 100 ms, logical zero is written in the same memory cell. In the last 100 ms, a reading impulse with a low level is applied to the corresponding memristor [40]. The write- and read-enable signals applied to the gate electrodes of the transistors T_1 and T_3 are also impulse signals with different magnitudes and polarities. For selecting the memristor element M_1, positive potentials with a value of 2 V are applied to the gate electrodes of both MOS transistors T_1 and T_3 with respect to their source electrodes. The resistance of the memristor element changes in a very large range [40]. Practically, the memristor element operates in a regime close to hard-switching state.

After a comparison of the current–voltage relationships of the memristor cell derived by the use of the applied model [22,47] and the improved window function [35] with experimentally recorded current–voltage characteristics [21] acquired under similar conditions, a reasonably good similarity between them is established.

4.5.3. Results and Discussion

The results acquired by the use of the applied model are identical to the experimentally recorded relationships [21].

The time graphs of the state variable, the resistance of the memristor element M_1 and the output voltage derived after a reading process by low-level pulses applied to the source of T_1 are presented in Figure 128(a–c) for explaining the memory device operation [40]. These time graphs illustrate the variation range of the resistance of the memristor element and the result derived by the applied reading signal [40]. It is clear that the state variable and the resistance of the corresponding memory element change in a very large range [40]. The wide alteration range of the memristance of the memory cell is useful for the correct recognition of the logical signals.

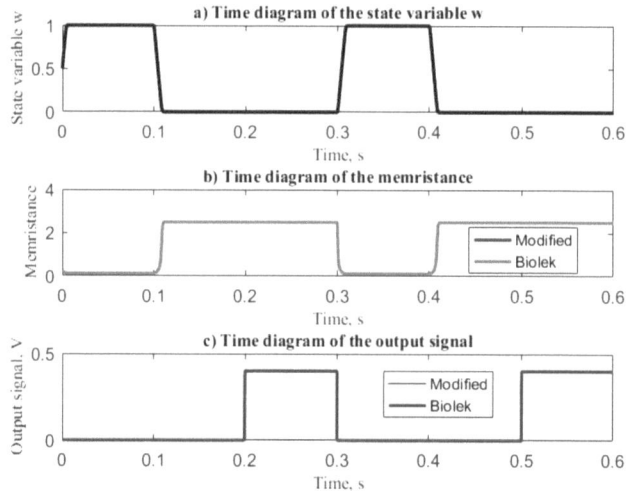

Figure 128. (a) Time graph of the state variable x; (b) Time diagrams of the memristance of the element M_1 according to the modified model and Biolek model; (c) Time diagram of the output voltage signal derived after reading the memory fragment, using the applied memristor model, with the modified and the standard Biolek window functions, presented for confirmation of the correct work of the memristor memory scheme.

The output voltage signal of the memory circuit, derived during the reading processes, has a reasonably high value, which is a good precondition for a correct separation of the logical signals [41]. Comparing the current–voltage relationships of the memristor element with experimental data [20,21] derived under similar conditions, a good rough correspondence with respect to the ranges of current and voltage and the current–voltage relationship form is established [40]. Identical results are acquired and it could be established that the proposed nonlinear memristor model with an improved window function could be applied for investigation of many complex memristor electronic schemes and devices [40].

An advantage of the suggested model [21,40] with the improved window function [35] applied here, compared to several existing models such as GBCM [14], Biolek [6] and Joglekar [5] models, is the reasonably high nonlinearity of the ionic dopant drift of the applied memristor model and the possibility for realistic illustration of the highly nonlinear relationship between the ionic current and the memristor voltage [35,40]. The sufficient convergence of the computing procedures using the corresponding algorithm for computer simulations is also a benefit of the applied model and the altered window function proposed by the author [35]. The successful operation of a fragment of resistance-switching memory circuit with the applied model and the improved window function for the general impulse mode confirms that this memristor model, together with the altered window function, could be applied for analysis of many memristor-based electronic schemes and devices. After investigation of the results acquired by the use of the applied memristor model [20,21] with the changed window function [35] and the standard Biolek window function [6], it could be established that they have almost similar performance based on the analysis of the memory operation processes: writing and reading information. The procedures of writing and reading in a hybrid memory fragment are successfully investigated with the suggested memristor model and the improved window function [35]. A good similarity between the derived and the experimentally recorded results [21] is established. The current–voltage relationship curves of the memristor element obtained by the suggested model [20,21] with the altered window function [35,40] and the standard Biolek window [6] are almost identical with respect to the ranges of current and voltage and current–voltage relationship outline. It could be established that the memristor model applied here with the altered window function [35] has the ability to illustrate the behavior of a memristor in memory devices for different electric regimes. An advantage of the suggested model [21,25] and the modified window [35], compared to the reference Pickett memristor model, is the established absence of computational issues [40]. Another advantage of the used memristor model is the use of a strongly nonlinear window function [35,40]. The applied memristor model with the improved window function [35] is suitable for realistic representation of the performance of the memristors in electronic devices and schemes for high voltages without convergence problems [40]. The suggested applied memristor model with the altered window function could be used for depiction of the memristor behavior for both soft-switching and hard-switching states [40].

CONCLUSION

In the present monograph, the basic analyses of memristors and memristor circuits and networks are described. A detailed reference check for memristor-based research is completed. The titanium dioxide memristors are still very applicable to many areas, and hence they are extensively investigated in the present researches. Owing to several disadvantages of the existing memristor models observed in the references, the author proposes modifications to the existing memristor models. These modifications are mainly associated with the improvement of the state differential equation for several basic models—Joglekar and Biolek memristor models. For the improvement of the relationship between the time derivative of the memristor state variable and the current, the author proposes two main variants for modification of the memristor state differential equation.

The first modification is proposing a voltage-dependent positive integer exponent in the applied memristor window function. This relationship relates the integer exponent applied to the used window function to the memristor voltage with a hyperbolic-like decreasing function. The reason for introducing this relationship is the established dependence between the nonlinearity of the ionic dopant drift and the applied memristor voltage. By increasing the voltage, the nonlinearity of the drift of the oxygen vacancies increases and it becomes an approximated exponential function of the voltage. On the other hand, the window function nonlinearity increases with the decrease of the applied positive integer exponent. Then, it could be concluded that the increase of the ionic drift nonlinearity could be realistically modeled by decreasing the integer exponent in the applied window function. In other words, for realistic representation of the nonlinear ionic dopant drift, a nonlinear decreasing dependence between the window function exponent and the memristor voltage is needed to be introduced in the corresponding state differential equation. The improved models have the capability of automatically changing their integer exponents in the window function in accordance to the applied memristor voltage. The basic advantage of the suggested models is the realistic representation of the memristor behavior in electric fields. The introduced nonlinearity of the models by the author is related to the memristor state variable. The proposed memristor modified models by the author have been investigated and the derived results have been compared with experimentally recorded characteristics of real memristor devices. A good similarity between the respective memristor characteristics derived with the suggested models and the experimental results is established. A simple linear combination of Joglekar and Biolek window functions with a voltage-dependent integer exponent is also applied, with a combination of a physical memristor model, for realistic representation of the memristor element behavior for the general electric mode.

The second basic modification of the memristor models is related to the application of an additional sinusoidal window function component in the standard

Biolek window function. For the alteration of the suggested model, a weighted sinusoidal window function component is used. The derived nonlinearity of the modified memristor model is higher than that of the standard Biolek model. In this modified model, an activation threshold is also applied. If the weight coefficient in front of the applied additional sinusoidal component of the window function and the corresponding activation threshold are zero, then the modified model is transformed into the standard Biolek model, so it could be concluded that the standard Biolek memristor model is a special case of the modified model by the author. The use of an activation threshold allows analyzing a memristor element for very-low-voltage signals. Then, the memristor element behaves as a linear resistor. The described properties of the modified memristor model are advantageous over those of the standard Biolek and Joglekar models.

Several memristor-based devices, such as series and parallel circuits, a generator, memory crossbars and artificial neurons, are investigated with the use of the proposed memristor models. The acquired results are compared with experimental data derived by physical measurements, and a good similarity of their current–voltage relationships is established. It could be concluded that the modified memristor models could be used for analysis of many different memristor-based electronic circuits, devices and networks.

The problems, following the main purpose of the present research, are completely executed. The analyses are mainly made by using a numerical solution to the basic system of equations of the memristor element solved with the finite difference method. The relationships between the ionic mobility of the oxygen vacancies and the specific resistance of the titanium dioxide memristor and the temperature are analyzed and expressed by a high-order polynomial. The dependence between the internal diffusion intensity and the temperature is also investigated. It is established that for lower temperatures, the operation of the memristor in electric fields is stable and improved, according to the diffusion processes between the doped and the undoped regions.

REFERENCES

1. Dearnaley, G.; Stoneham, A.; Morgan, D. Electrical phenomena in amorphous oxide films. *Rep. Prog. Phys.* **1970**, *33*, 1129–1191.

2. Jeong, D.S.; Schroeder, H.; Waser, R. Coexistence of bipolar and unipolar resistive switching behaviors in a $Pt/TiO_2/Pt$ stack. *Electrochem. Solid State Lett.* **2007**, *10*, G51–G53.

3. Chua, L. Memristor—The Missing Circuit Element. *IEEE Trans. Circuit Theory* **1971**, *18*, 507–519.

4. Strukov, D.; Snider, G.; Stewart, D.; Williams, R.S. The Missing Memristor Found. *Nat. Lett.* **2008**, *453*, 80–83. [CrossRef]

5. Joglekar, Y.; Wolf, S. The Elusive Memristor: Properties of Basic Electrical Circuits. *Eur. J. Phys.* **2009**, *30*, 661–675. [CrossRef]

6. Biolek, Z.; Biolek, D.; Biolkova, V. SPICE Model of Memristor with Nonlinear Dopant Drift. *Radioengineering* **2009**, *18*, 210–214.

7. Abdalla, H.; Pickett, M. SPICE modelling of memristors. *IEEE Int. Symp. Circuits Syst.* **2011**, 1832–1835. [CrossRef]

8. Chen, Y.; Liu, G.; Wang, C.; Zhang, W.; Li, R.; Wanga, L. Polymer memristor for information storage and neuromorphic applications. *Mater. Horiz.* **2014**, *1*, 489–506. [CrossRef]

9. Hu, Z.; Li, Q.; Li, M.; Wang, Q.; Zhu, Y.; Liu, X.; Zhao, X.; Liu, Y.; Dong, S. Ferroelectric memristor based on $Pt/BiFeO_3/Nb$-doped $SrTiO_3$ heterostructure. *Appl. Phys. Lett.* **2013**, *102*, 1–5.

10. Mostafa, H.; Ismail, Y. Process Variation Aware Design of Multi-Valued Spintronic Memristor-Based Memory Arrays. *IEEE Trans. Semicond. Manuf.* **2016**, *29*, 145–152. [CrossRef]

11. Corinto, F.; Ascoli, A. A Boundary Condition-Based Approach to the Modeling of Memristor Nanostructures. *IEEE Trans. Circuits Syst.* **2012**, *59*, 2713–2726. [CrossRef]

12. Mladenov, V.; Vladov, S. *Theory of Electrical Engineering*; KING Publishing House: Sofia, Bulgaria, 2013; ISBN 978-954-9518-74-0.

13. Palm, W. *Introduction to MATLAB for Engineers*; McGraw-Hill: New York, NY, USA, 2011; ISBN 978-1-259-01205-1.

14. Ascoli, A.; Corinto, F.; Tetzlaff, R. Generalized Boundary Condition Memristor Model. *Int. J. Circ. Theor. Appl.* **2016**, *44*, 60–84. [CrossRef]

15. Ascoli, A.; Tetzlaff, R.; Corinto, F.; Gilli, M. PSpice switch-based versatile memristor model. *IEEEInt. Symp. Circuits Syst.* **2013**, 205–208. [CrossRef]

16. Walsh, A.; Carley, R.; Feely, O.; Ascoli, A. Memristor circuit investigation through a new tutorial toolbox. *IEEE ECCTD Germany* **2013**, 1–4. [CrossRef]

17. Pickett, M.; Strukov, D.; Borghetti, J.; Yang, J.; Snider, G.; Stewart, D.; Williams, R.S. Switching dynamics in titanium dioxide memristive devices. *J. Appl. Phys.* **2009**, *106*, 1–6. [CrossRef]

18. Simmons, J. Electric tunnel effect between dissimilar electrodes separated by a thin insulating film. *J. Appl. Phys.* **1963**, *34*, 2581–2590.

19. Rashid, M. *Introduction to PSpice Using OrCAD for Circuits and Electronics*, 3rd ed.; Prentice Hall: Upper Saddle River, NJ, USA, 2004; ISBN 0-13-101988-0.

20. Yang, J.J.; Pickett, M.D.; Li, X.; Ohlberg, D.A.A.; Stewart, D.R.; Williams, R.S. Memristive switching mechanism for metal/oxide/metal nanodevices. *Nat. Nanotechnol.* **2008**, *3*, 429–433.

21. Lehtonen, E.; Laiho, M. CNN using memristors for neighborhood connections. 2010 12th International Workshop on Cellular Nanoscale Networks and Their Applications (CNNA), Berkeley, CA, USA, 3–5 February 2010; pp. 1–4.

22. Ascoli, A.; Corinto, F.; Senger, V.; Tetzlaff, R. Memristor Model Comparison. *IEEE Circuits Syst. Mag.* **2013**, 89–105. [CrossRef]

23. Ascoli, A.; Tetzlaff, R.; Biolek, Z.; Kolka, Z.; Biolkovà, V.; Biolek, D. The Art of Finding Accurate Memristor Model Solutions. *IEEE J. Emerg. Sel. Top. Circuits Syst.* **2015**, *5*, 133–142. [CrossRef]

24. Linn, E.; Siemon, A.; Waser, R.; Menzel, S. Applicability of Well-Established Memristive Models for Simulations of Resistive Switching Devices. *IEEE Trans. Circuits Syst.* **2014**, *61*, 2402–2410.

25. Jiang, Z.; Wu, Y.; Yu, S.; Yang, L.; Song, K.; Karim, Z.; Wong, H.-P. A Compact Model for Metal-Oxide Resistive Random Access Memory with Experiment Verification. *IEEE Trans. Electron. Devices* **2016**, *63*, 1884–1892.

26. Bak, T.; Nowotny, J.; Rekas, M.; Sorrell, C.C. Defect chemistry and semiconducting properties of titanium dioxide: III. Mobility of electronic charge carriers. *J. Phys. Chem. Solids* **2013**, *64*, 1069–1087.

27. Mladenov, V.; Kirilov, S. Analysis of temperature influence on titanium dioxide memristor characteristics at pulse mode. In Proceedings of the ISTET 2013: International Symposium on Theoretical Electrical Engineering, Pilsen, Czech Republic, 24–26 June 2013; pp. 1–6, ISBN 978-80-261-0246-5.

28. Kim, M.; Baek, S.; Paik, U. Electrical Conductivity and Oxygen Diffusion in Nonstochiometric TiO_{2-x}. *J. Korean Phys. Soc.* **1998**, *32*, 1127–1130.

29. Cotton, A.F.; Wilkinson, G. *Advanced Inorganic Chemistry*; A Wiley Interscience Publication: New York, NY, USA, 1980; ISBN 0-471-02775-8.

30. Mehrer, H. *Diffusion in Solids*; Springer: Berlin, Germany, 2007.

31. Mladenov, V.; Kirilov, S. Investigation of memristors' own parasitic parameters and mutual inductances between neighboring elements of a memristor matrix and their influence on the characteristics. In Proceedings of the ISTET 2013: International Symposium on Theoretical Electrical Engineering, Pilsen, Czech Republic, 24–26 June 2013; pp. 1–6, ISBN 978-80-261-0246-5.

32. Mladenov, V.; Kirilov, S. A Nonlinear Memristor Model with Activation Thresholds and Variable Window Functions. In Proceedings of the IEEE Proceedings of CNNA 2016; 15th International Workshop on Cellular Nanoscale Networks and their Applications, Dresden, Germany, 23–25 August 2016; pp. 1–2, ISBN 978-3-8007-4252-3.

33. Mladenov, V.; Kirilov, S. A Nonlinear Drift Memristor Model with a Modified Biolek Window Function and Activation Threshold. *Electronics* **2017**, *6*, 77. [CrossRef]

34. Mladenov, V.; Kirilov, S. Advanced Memristor Model with a Modified Biolek Window and a Voltage-Dependent Variable Exponent. *Informatyka, Automatyka, Pomiary w Gospodarce i Ochronie Środowiska (IAPGOS)* **2018**, *8*, 15–20. [CrossRef]

35. Mladenov, V.; Kirilov, S. A Memristor Model with a Modified Window Function and Activation Thresholds. In Proceedings of the IEEE ISCAS 2018, Florence, Italy, 27–30 May 2018; pp. 1–5. [CrossRef]

36. Kirilov, S.; Mladenov, V. Integrator Device with a Memristor Element. In Proceedings of the 2018 IEEE, 7th International Conference on Modern Circuits and Systems Technologies (MOCAST), Thessaloniki, Greece, 7–9 May 2018; pp. 1–4. [CrossRef]

37. Mladenov, V. Synthesis and Analysis of a Memristor-Based Artificial Neuron. In Proceedings of the 16th International Workshop on Cellular Nanoscale Networks and their Applications (CNNA) 2018, Budapest, Hungary; ISBN 978-3-8007-4766-5.

38. Kirilov, S.; Mladenov, V. Analysis of a Passive Memristor Crossbar. *Orient. J. Comput. Sci. Technol.* **2018**, *11*, 4–11. [CrossRef]

39. Kirilov, S.; Mladenov, V. Learning of an Artificial Neuron with Resistor-Memristor Synapses. In Proceedings of the workshop on Advances in Neural Networks and Applications 2018 (ANNA'18), 2018 VDE VERLAG GMBH, Berlin, Offenbach; pp. 13–17, ISBN 978-3-8007-4756-6.

40. Mladenov, V. Analysis and Simulations of Hybrid Memory Scheme Based on Memristors. *Electronics* **2018**, *7*(11), 289. [CrossRef]

41. Mladenov, V.; Kirilov, S. Analysis of a serial circuit with two memristors and voltage source at sine and impulse regime. In Proceedings of the IEEE, 13th International Workshop on Cellular Nanoscale Networks and Their Applications (CNNA), Turin, Italy, 29–31 August 2012; pp. 1–6. [CrossRef]

42. Mladenov, V.; Kirilov, S. Syntheses of a PSPICE model of a titanium-dioxide memristor and Wien memristor generator. In Proceedings of the IEEE, 2013 European Conference on Circuit Theory and Design (ECCTD), Dresden, Germany, 8–12 September 2013; pp. 1–4. [CrossRef]

43. Mladenov, V.; Kirilov, S. Analysis of an anti-parallel memristor circuit. *Informatyka, Automatyka, Pomiary w Gospodarce i Ochronie Środowiska (IAPGOS)* **2018**, *8*, 9–14. [CrossRef]

44. Fausett, L. *Fundamentals of Neural Networks*; Prentice Hall: Englewood Cliffs, NJ, USA, 1994; ISBN 0130422509.

45. Sah, M.; Yang, C.; Kim, H.; Roska, T.; Chua, L. Memristor Bridge Circuit for Neural Synaptic Weighting. In Proceedings of the 13th International Workshop on Cellular Nanoscale Networks and Their Applications (CNNA), IEEE, Turin, Italy, 29–31 August 2012; pp. 1–5. [CrossRef]

46. Mladenov, V.; Kirilov, S. Synthesis and Analysis of a Memristor-Based Perceptron for Logical Function Emulation. *Przegląd Elektrotechniczny* **2016**, 22–25. [CrossRef]

47. Haron, N.; Hamdioui, S. On Defect Oriented Testing for Hybrid CMOS/Memristor Memory. *IEEE Asian Test Symp.* **2011**, 353–358. [CrossRef]

48. Serb, A.; Redman-White, W.; Papavassiliou, C.; Prodromakis, T. Practical determination of individual element resistive states in selectorless RRAM arrays. *IEEE Trans. Circuits and Syst I* **2016**, *63*, 827–835. [CrossRef]
49. Strukov, D.B.; Williams, R.S. Exponential ionic drift: Fast switching and low volatility of thin-film memristors. *Appl. Phys. A, Mater. Sci. Process.* **2009**, *94*, 515–519.

MDPI

St. Alban-Anlage 66

4052 Basel

Switzerland

Tel. +41 61 683 77 34

Fax +41 61 302 89 18

www.mdpi.com

MDPI Books Editorial Office

E-mail: books@mdpi.com

www.mdpi.com/books

MDPI

www.ingramcontent.com/pod-product-compliance
Lightning Source LLC
Chambersburg PA
CBHW051559190326
41458CB00029B/6480